Practical Approach to
Crystallography and Mineralogy

Second Edition

W0193382

Practical Approach to
Crystallography and Mineralogy

Second Edition

Rabindra Nath Hota

Professor
PG Department of Geology
Utkal University
Bhubaneswar
Odisha

CBSPD

CBS Publishers & Distributors Pvt Ltd

New Delhi • Bengaluru • Chennai • Kochi • Kolkata • Lucknow • Mumbai
Hyderabad • Jharkhand • Nagpur • Patna • Pune • Uttarakhand

Practical Approach to

Crystallography and Mineralogy

Second Edition

ISBN: 978-93-86217-69-1

Copyright © Author and Publisher

Second Edition: 2017

Reprint: 2023

First Edition: 2011

Reprint: 2012

Published by Satish Kumar Jain and produced by Varun Jain for

CBS Publishers & Distributors Pvt Ltd

4819/XI Prahlad Street, 24 Ansari Road, Daryaganj, New Delhi 110 002, India
Ph: 011-23289259, 23266861, 23266867 Fax: 011-23243014   Website: www.cbspd.com
e-mail: delhi@cbspd.com; cbspubs@airtelmail.in.

Corporate Office: 204 FIE, Industrial Area, Patparganj, Delhi 110 092, India
Ph: 011-4934 4934   Fax: 011-4934 4935   e-mail: publishing@cbspd.com; publicity@cbspd.com

Branches

- **Bengaluru:** Seema House 2975, 17th Cross, KR Road, Banasankari 2nd Stage, Bengaluru 560 070, Karnataka, India
  Ph: +91-80-26771678/79   Fax: +91-80-26771680   e-mail: bangalore@cbspd.com
- **Chennai:** 7, Subbaraya Street, Shenoy Nagar, Chennai 600 030, Tamil Nadu, India
  Ph: +91-44-26680620, 26681266   Fax: +91-44-42032115   e-mail: chennai@cbspd.com
- **Kochi:** 42/1325, 1326, Power House Road, Opp KSEB, Power House, Ernakulam 682 018, Kerala, India
  Ph: +91-484-4059061-65/67   Fax: +91-484-4059065   e-mail: kochi@cbspd.com
- **Kolkata:** 147, Hind Ceramics Compound, 1st Floor, Nilgunj Road, Belghoria, Kolkata 700 056, West Bengal, India
  Ph: +033-25633055, 033-25633056   e-mail: kolkata@cbspd.com
- **Lucknow:** Basement, Khushnuma Complex, 7-Meerabai Marg (Behind Jawahar Bhawan), Lucknow 226 001, UP, India
  Ph: 0522-4000032   e-mail: tiwari.lucknow@cbspd.com
- **Mumbai:** PWD Shed. Gala no. 25/26, Ramchandra Bhatt Marg, Next to JJ Hospital Gate no. 2, Opp. Union Bank of India, Noorbaug Mumbai 400 009, Maharashtra, India
  Ph: 022-66661880/89   e-mail: mumbai@cbspd.com

Representatives

Hyderabad	0-9885175004	**Jharkhand**	0-9811541605	**Nagpur**	0-9421945513
Patna	0-9334159340	**Pune**	0-9623451994	**Uttarakhand**	0-9716462459

Printed at Glorious Printers, Delhi, India

Foreword

The number of books on practical aspects of geology is meager. This problem gets compounded for the students due to lack of good books on practical aspects of geology. Some of the books which are available in the market do not go much deeper into theoretical aspects due to constraints of either space or financial matter. This makes those books somewhat out of place.

This endeavour by Dr RN Hota is truly praiseworthy. Part I of this volume which deals with crystallography gives in-depth information on classification of crystals into different systems and classes as per the latest classification schemes, stereographic projection of crystals, determination of axial ratio, and problems related to zonal laws. The portion on mineralogy (Part II) dealing with physical and optical properties of minerals in both transmitted and reflected light is a valiant attempt by the author to cover all the determinative aspects of all groups of minerals. Determination of chemical formula from the chemical composition is a welcome addition.

I am sure the book will cater to the needs of the geology students of higher secondary, undergraduate and postgraduate classes. I am overwhelmed with joy to endorse this work of Dr Hota to the geosciences fraternity.

Dr HK Sahoo
Professor and Head
PG Department of Geology
Utkal University
Bhubaneswar
Odisha

Preface to the Second Edition

The book *Practical Approach to Crystallography and Mineralogy* published in 2011 January got reprinted in 2012, which signifies the overwhelming response from the readers, particularly the students for whom it has been written. I thank all the readers for their patronage. In this edition, the stray typographical and other errors of the first edition have been eradicated. In addition, many of the paragraphs have been rewritten and rearranged to maintain homogeneity of the chapters and also for better readability. Examples of identification and stereographic projection of hexagonal combination form have been added to make the book complete.

I sincerely appeal the readers (teachers and students) to suggest for betterment of this book in future. They are requested to send their feedbacks and suggestions to me through e-mail (*rnhota@yahoo.com*).

Rabindra Nath Hota

Preface to the First Edition

The thought of writing a textbook in geology crept into my mind in 1987 when I started my teaching career as a lecturer in the undergraduate section of the Department of Geology, Utkal University, located at Ravenshaw College, Cuttack. I could realise the problems faced by students who join the degree classes directly to make geology as their career. Further, shortage of teachers and supporting staff makes the problem more acute. I started writing after completing my PhD and postdoctoral research works.

Like other science subjects, both theory and practical are equally important in geology in addition to the field study. A few textbooks and a number of reference books on the theory part of the subject are available in the Indian market. Realising the difficulties faced by the students in practical classes, I started writing textbook on practical aspects of different branches in geology keeping in mind the revised curriculum recommended by University Grants Commission.

Mineralogy and crystallography are the fundamental branches of geology and are prerequisites to study different types of rocks. So I started with writing practical approach to crystallography and mineralogy. The crystallography part of this book embodies crystal morphology, study and identification of crystals of all the systems, stereographic projection including determination of axial ratio and zone and zonal laws. The mineralogy part includes study and identification of minerals as megascopic specimens and in thin sections including determination of mineral formula from analytical data. Some of the sections have been illustrated with numerical problems and some more problems have been added for practice.

In the process of writing, a number of books written by different authors have been followed including incorporation of handouts provided to the students in practical classes. Any part of the text, method, statement and illustrations incorporated in the writing of this book is acknowledged. I am indebted and thankful to the authors and publishers of those referred materials. I am very much thankful to Prof Wataru Maejima of Osaka City University, Japan, who laid the foundation stone of this book writing by teaching me the methods of drawing/redrawing of figures. Due to his kind help, this book has been illustrated with a lot of figures. I express my hearty gratitude to Prof HK Sahoo of PG Department of Geology, Utkal University, who read major part of the manuscript, gave valuable suggestions to make the book complete in all respects and kindly agreed to write the Foreword to this book. However, I accept the responsibility of any mistakes which might have crept in inadvertently. In spite of the critical scrutiny, some typographical errors might have been left out. I shall be very much thankful to the readers if lapses of any kind are communicated to me (rnhota@yahoo.com).

I express my gratitude to Prof (Mrs) M Das, Dr PP Singh, Dr BK Ratha, Mr MR Mohapatra and Mr G Das of the PG Department of Geology, Utkal University, for their encouragement and help during preparation of the manuscript.

I am thankful to Mr YN Arjuna, Senior Director—Publishing, Editorial and Publicity, and M/s CBS Publishers & Distributors, New Delhi, for their keen interest in publishing the book.

Last but not the least, I thank my wife Anjali and children Swayam Prakash and Soumya Ranjan for their inspiration and cooperation during preparation of the manuscript.

(Rabindra Nath Hota)

Contents

Crystal Morphology

Crystallography deals with the study of crystals. It includes the description of crystals, their classification into different systems and classes in addition to the study of mathematical relationship between different faces. Study of imperfections of crystal growth and crystalline aggregates are also included in the domain of crystallography.

A crystal is a regular polyhedral form characterised by flat surfaces known as faces. The development of faces is the manifestation of the internal ordered arrangement of the atoms and depends on the physicochemical conditions that prevail at the time of formation of the mineral by solidification of gas or liquid or precipitation from solution. Thus, a crystal is a special form of a mineral.

1.1 EXTERNAL MORPHOLOGY OF CRYSTAL

The most important attribute of a crystal is the presence of flat faces and their arrangement in a distinctive pattern that results in the development of some sort of symmetry. The common faces are normally parallel to net-planes containing maximum number of lattice points. A crystal may have like faces, which are similar in shape and size as in Fig. 1.1, where all the faces are triangular of equal size and shape. Such a crystal is said to be a simple form. A crystal may be bounded by dissimilar or unlike faces as in Fig. 1.2, where two types of faces are present, one is trapezoid and the other is triangular. In this case, the crystal is said to be a combination form. In addition, a crystal may be termed a closed form (solid) when it encloses some space and open form when

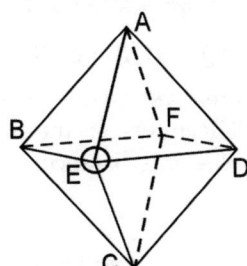

Fig. 1.1: Simple form showing a solid angle at E

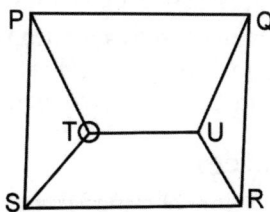

Fig. 1.2: Combination form showing a solid angle at T

it does not enclose any space. Thus, a closed form crystal can occur independently, whereas an open form occurs in association with other form(s), giving rise to a combination form. In naturally occurring combination forms, two or more sets of faces may be present. Due to paucity of naturally occurring crystals, models made up of wood, plastic, glass or any other materials are generally used in practical classes.

An edge is formed by the intersection of two adjacent faces. It is parallel with the row of atoms occurring at the intersection of net-planes. In Fig. 1.1, AB, BC, CD etc are edges. A solid angle is formed by the intersection of three or more faces. In Fig. 1.1 the circle at 'E' indicates a solid angle formed by the intersection of faces ABE, BCE, CDE and AED. Similarly, in Fig. 1.2 the circle at T indicates a solid angle formed by the intersection of three faces PQUT, PTS and STUR.

1.2 REPRESENTATION OF CRYSTAL FACE

Since a crystal is a three-dimensional polyhedral form, a set of reference axes similar to those of solid geometry is used to describe the character of its constituent faces. However, the axes used in crystallography are of definite length and are differently oriented unlike those of solid geometry. The character and nature of the faces are expressed by intercepts, parameters and indices.

1.2.1 Intercepts

Intercepts are the distances between the centre of the crystal (point of intersection of the crystallographic axes) and the points of intersection of the face and axes. In Fig. 1.3, OX, OY and OZ axes are denoted by conventional crystallographic symbols a, b and c respectively. ABC and PQR are two crystal faces. The intercepts of the face ABC on a, b and c axes are OA, OB and OC respectively. Similarly the intercepts of the face PQR are OP, OQ and OR respectively.

1.2.2 Parameters

Parameters are the ratios of the intercepts, i.e. the intercepts of the face in consideration are compared with those of the unit form. If ABC is a face of the unit form, then OA, OB and OC are unit distances in a, b and c axes respectively.

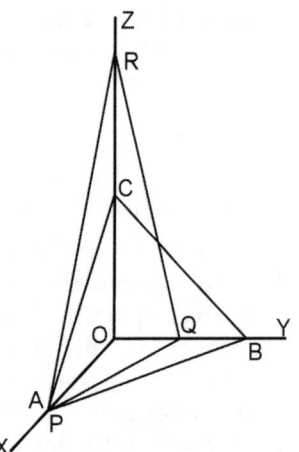

Fig. 1.3: Crystallographic axes with two faces ABC and PQR

Parameters of the face PQR are: OP/OA, OQ/OB and OR/OC on a, b and c axes respectively.

This is expressed as 1a, ½ b and 2c   [Since OP = OA, OQ = ½ OB and OR = 2OC]

This type of representation of parameters is known as parameter system of Weiss. The symbols 1a, ½ b and 2c indicate that the face under consideration (PQR) intersects the a-axis at unit distance, b-axis at half of the unit distance and c-axis at twice of the unit distance.

1.2.3 Indices

The indices are obtained from the Weiss parameters by taking the reciprocals and clearing the fractions. The indices are written in axial order, i.e. a, b and c. Hence, the letters a, b and c associated with the parameters are omitted. The indices of the face PQR are

$$\frac{1}{1}\,\frac{1}{\frac{1}{2}}\,\frac{1}{2} \text{ or } 12\frac{1}{2} \text{ or } 241$$

This system of representation of indices is known as index system of Miller. For a face that intersects the a-axis at unit distance, b-axis at thrice of the unit distance and remains parallel with the c-axis, the Weiss parameters are 1a, 3b, ∝c (It is assumed that the face intersects the c-axis at infinite distance). The miller indices are $\frac{1}{1}\,\frac{1}{3}\,\frac{1}{\infty}$ or 310 (read as three, one, naught but not as three one zero or three hundred ten). A face may intersect all the three axes at unequal lengths. In such cases the symbols may be 123, 321, 234 or any other numerical depending on the lengths of the intercepts. If the lengths of intercepts cannot be determined decisively the general symbol 'hkl' may be used. The intercepts and indices are inversely related, i.e. the highest intercept corresponds to the lowest index and vice versa. The Miller indices of a face are also known as the symbol of the corresponding face.

1.3 REPRESENTATION OF FORM SYMBOL

A simple form consists of a set of fixed number of faces which has similar characteristics, i.e. shape, size and crystallographic orientation. For example, a cube (Fig. 1.4) consists of six square faces each of which intersects one crystallographic axis at unit distance and remains parallel with other two axes.

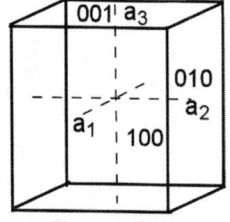

Fig. 1.4: Cube

The Weiss parameters of the face towards the observer that intersects the a_1-axis and remains parallel with other two axes are 1∝∝ and the corresponding Miller indices are 100. The symbols of faces to the right and top are 010 and 001 respectively. The symbols of remaining three faces, which intersect the axes on the negative sides are $\overline{1}00$, $0\overline{1}0$ and $00\overline{1}$. Out of these six symbols, one is to be used to represent the form, i.e. the cube. There is disagreement among various authors regarding representation of face and form symbols. Dana and Ford as well as Read indicate the face symbol without bracket and form symbol within '( )' bracket. Hurlbut and Klein, on the other hand, prefer to keep face symbol within '( )' bracket and form symbol within '{ }' bracket. Generally the symbol of one of the faces of a crystal is taken to represent the form symbol. There is also lack of unanimity regarding selection of the form symbol. Dana and Ford as well as Read prefer to follow the convention h>k>l, i.e. the first digit of the symbol is the greatest and last digit of the symbol is the least. Thus, the cube is represented by the symbol (100). Hurlbut and Klein and others follow the convention h<k<l, i.e. the first digit of the symbol is the least and last digit of the symbol is the greatest. According to them, the symbol {001} represents the cube. In both cases, the form symbol does not indicate the number of faces present in the crystal. Thus, two or more forms having different number of faces but similar relationship with the crystallographic axes are represented by the same symbol. To avoid these confusions, in this book, the general symbol of the form that intersects all the three crystallographic axes is represented by the symbol '(hkl)$_n$' where generally h<k<l and 'n' is the number of faces present in the form. Thus, the cube is represented by the symbol (001)$_6$, which indicates that it is a form having six faces all of which intersect one crystallographic axis at unit distance and remain parallel with other two axes. Hence, it is a unique symbol that differs from the form symbols like (001)$_2$ (pinacoid) and (001)$_1$(pedion), which have dissimilar number of faces but similar crystallographic relationship.

1.4 CRYSTAL SYMMETRY

Regularity in the arrangement of like faces, edges, and solid angles is the characteristic feature of a crystal. This regularity constitutes the symmetry of the crystal and depends on the internal atomic structure of the mineral. The symmetry can be defined with reference to three criteria, viz. plane of symmetry, axis of symmetry and centre of symmetry generated by reflection, rotation and inversion operations respectively.

1.4.1 Plane of Symmetry

A plane of symmetry divides a figure or a crystal into two similar and similarly placed halves, which are mirror images of each other. It appears as if one-half of the figure or the crystal is generated by reflection over a plane. The plane of symmetry is also known as *mirror plane* or *mirror* and is denoted by the letter 'm' ('P' in some books). The presence of the plane of symmetry can be well visualized from different two-dimensional figures shown in Figs 1.5a to g. The dashed lines represent planes of symmetry. Figure 1.5a is an irregular figure which is not symmetrical with respect to any plane and thus, lacks a plane of symmetry. The isosceles triangle, rectangle, equilateral triangle, square and hexagon shown in Figs 1.5b, c, d, e and f have 1, 2, 3, 4 and 6 number of planes of symmetry respectively shown by dashed lines. The circle (Fig.1.5g) has infinite number of planes of symmetry (only a few are shown) each coincident with a diameter of the circle. Two three-dimensional crystal forms, a prism and a cube, are shown in Figs 1.6 and 1.7 respectively. The upper two faces of the prism are related with the lower two faces by a horizontal plane of symmetry. Similarly, a cube has nine planes of symmetry indicated by numbers '1' to '9'. A sphere has infinite number of planes of symmetry.

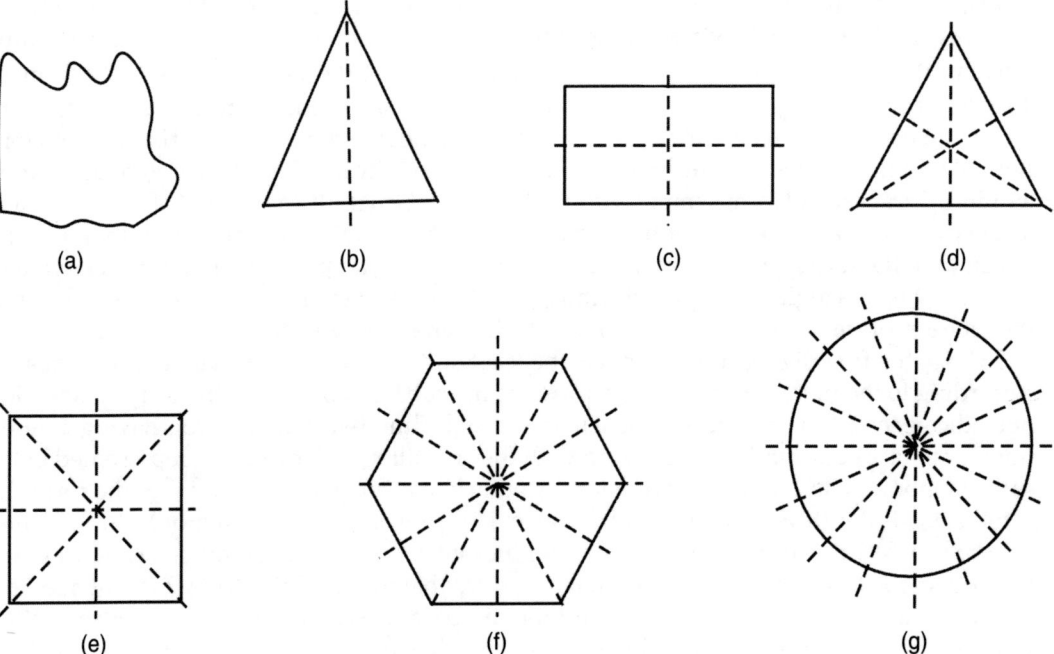

(a) (b) (c) (d)

(e) (f) (g)

Fig.1.5: Planes of symmetry

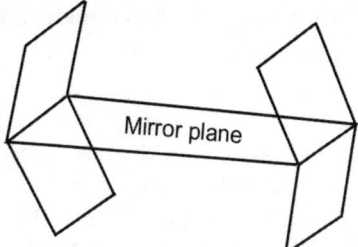

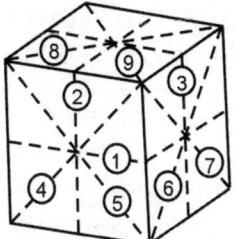

Fig. 1.6: Mirror plane in a prism

Fig. 1.7: Nine symmetry planes in cube

1.4.2 Axis of Symmetry

Symmetry operations like rotation and rotation combined with inversion give rise to axes of symmetry. The rotational axis, also known as simple *axis of symmetry*, involves rotation of the crystal about an imaginary line passing through the centre of the crystal about which if the crystal is rotated by certain angle occupies the immediately next similar position in space. The angle between two successive similar positions in space is known as the *elementary angle of rotation*. The axes of symmetry are 1-fold, 2-fold, 3-fold, 4-fold and 6-fold with elementary angles of rotation 360°, 180°, 120°, 90° and 60° are represented by numbers 1, 2, 3, 4 and 6 respectively (Figs 1.8a to e). The axes of 2-fold, 3-fold, 4-fold and 6-fold are also known as diad, triad, tetrad and hexad axes respectively.

The degree of the axis indicates the number of times the crystal occupies the same position in space in a complete rotation of 360° about the concerned axis. Axis of 1-fold rotation (1) is universally present in all the objects whatever the shape may be and thus, is not regarded as a symmetry element by many authors. However, in the true sense, it indicates the lack of symmetry and is used in classification of crystals into 32 classes. A pinacoid shown in Fig. 1.8b consists of two faces, which interchange their positions twice in a complete rotation of 360°. In Figs 1.8c to e the trigonal, tetragonal and hexagonal pyramids possess rotational axes of 3-, 4- and 6-fold respectively. A form may have axes of different degrees. For example, a cube has six axes of 2-fold, four axes of 3-fold and three axes of 4-fold rotational symmetry.

The *rotoinversion axis* involves rotation of the crystal about it by elementary angle of rotation followed by inversion. Corresponding to five elementary angles mentioned above, there are five rotoinversion axes designated as $\overline{1}, \overline{2}, \overline{3}, \overline{4}$ and $\overline{6}$ read as *bar one, bar two, bar three, bar four* and *bar six*. $\overline{1}$ is equivalent to a centre of symmetry (or centre of inversion, i), $\overline{2}$ is equivalent to a

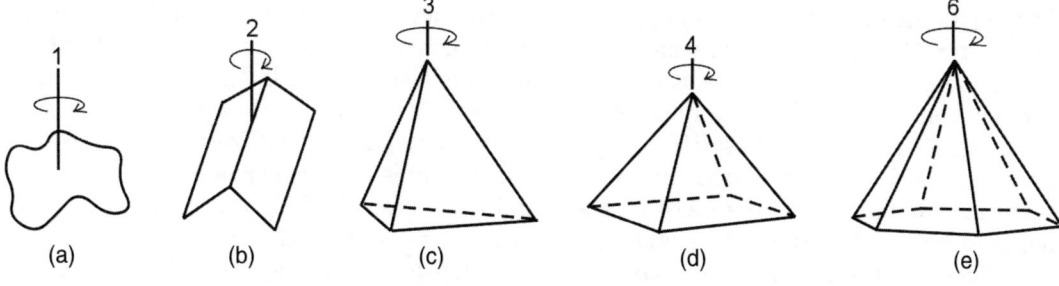

Fig. 1.8: Axes of symmetry

mirror plane (m), $\bar{3}$ is equivalent to a 3-fold rotation axis plus centre of symmetry, $\bar{4}$ is unique and $\bar{6}$ is equivalent to a 3-fold rotation axis with a perpendicular mirror plane.

1.4.3 Centre of Symmetry

A crystal is said to has a centre of symmetry if like faces, edges and solid angles are arranged in pairs in corresponding positions on opposite sides of a central point. It is observed that when a crystal having centre of symmetry is made to stand on a certain face, an identical face remains at the top of the crystal. A crystal having an axis of even-fold perpendicular to a mirror plane generally possesses a centre of symmetry.

1.5 CLASSIFICATION OF CRYSTALS INTO SYSTEMS

The crystals are classified into six systems each characterised by a set of crystallographic axes. The six systems are isometric, tetragonal, hexagonal, orthorhombic, monoclinic and triclinic.

The isometric system has three reference axes which are equal in length and perpendicular to each other and thus, are mutually interchangeable. The tetragonal system is characterised by the presence of three mutually perpendicular crystallographic axes out of which two horizontal axes are equal in length and the third axis is of different length and vertical in position. Crystals belonging to the hexagonal system are referred to four crystallographic axes, out of which three are of equal length; lie in the horizontal plane with angles of 120° between their positive ends. The fourth axis is vertical and is of different length. The orthorhombic system is characterised by three crystallographic axes, which are unequal in length but perpendicular to each other. The monoclinic system has three axes of unequal length; one vertical, another perpendicular to the vertical axis and the third is inclined to the plane containing the first two axes. The crystals of the triclinic system are referred to three axes of unequal length mutually inclined with each other at angles different from 90°.

Some authors classify the crystals into seven systems on the basis of crystal symmetry. In such a scheme, the hexagonal system is divided into two divisions, hexagonal proper, in which, the vertical crystallographic axis is an axis of 6-fold symmetry (rotation or inversion) and trigonal division, in which, the vertical crystallographic axis is an axis of 3-fold symmetry (rotation or inversion).

1.6 CLASSIFICATION OF CRYSTALS INTO 32 CLASSES

The six/seven systems mentioned above are divided into 32 classes each consisting of a group of forms. E. S. Dana and W. Ford, name these classes after the significant form(s) present in these classes. For example, they name the class in which calcite crystallizes as rhombohedral, ditrigonal scalenohedral, hexagonal scalenohedral, rhombohedral hemihedral and dihexagonal alternating. Though hexagonal scalenohedron is the general form of this class, familiarity of rhombohedron cannot be ruled out. At the same time, terms *hemihedral* and *alternating* have different connotations. Further, five names for a single class seems to be more than necessary. H. H. Read designates a class after a significant mineral that crystallizes in that class. For example, the hexoctahedral class of the isometric system is known as galena type. This also seems anomalous due to the fact that some highly precious and significant minerals like diamond, native platinum, gold, silver, copper, etc. also crystallise in this class. The term *normal class* refers to the class that shows maximum symmetry possible in a system. This name is not unique as each system has such a class. Thus, it is felt that a system should be followed in which each class will have a unique non-repetitive name characterised by a set of symmetry elements and thus, known as the *symmetry class*. The notation system developed by C. H. Hermann and C. V. Mauguin fulfils these criteria. The 32 classes with Hermann-Mauguin symbols, descriptive names and symmetry elements are given in Table 1.1.

Table 1.1: Thirty-two crystal classes with their symmetry elements

System	*HM class*	*Descriptive name*	*Symmetry element(s)*
Isometric	$4/m\,\overline{3}\,2/m$	Hexoctahedral	$i, 3A_4, 4\,\overline{A}_3, 6A_2, 9m$
	$\overline{4}\,3m$	Hextetrahedral	$3\,\overline{A}_4, 4A_3, 6m$
	432	Gyroidal	$3A_4, 4A_3, 6A_2$
	$2/m\,\overline{3}$	Diploidal	$i, 4\,\overline{A}_3, 3A_2, 3m$
	23	Tetartoidal	$4A_3, 3A_2$
Tetragonal	$4/m\,2/m\,2/m$	Ditetragonal dipyramidal	$i, 1A_4, 4A_2, 5m$
	$\overline{4}\,2m$	Tetragonal scalenohedral	$1\,\overline{A}_4, 2A_2, 2m$
	4 mm	Ditetragonal pyramidal	$1A_4, 4m$
	422	Tetragonal trapezohedral	$1A_4, 4A_2$
	$4/m$	Tetragonal dipyramidal	$i, 1A_4, 1m$
	$\overline{4}$	Tetragonal disphenoidal	$1\,\overline{A}_4$
	4	Tetragonal pyramidal	$1A_4$
Hexagonal	$6/m\,2/m\,2/m$	Dihexagonal dipyramidal	$i, 1A_6, 6A_2, 7m$
	$\overline{6}\,m2$	Ditrigonal dipyramidal	$1\,\overline{A}_6, 3A_2, 3m$
Hexagonal	6 mm	Dihexagonal pyramidal	$1A_6, 6m$
division	622	Hexagonal trapezohedral	$1A_6, 6A_2$
	$6/m$	Hexagonal dipyramidal	$i, 1A_6, 1m$
	$\overline{6}$	Trigonal dipyramidal	$1\,\overline{A}_6$
	6	Hexagonal pyramidal	$1A_6$
Rhombohedral/	$\overline{3}\,2/m$	Hexagonal scalenohedral	$i, 1\,\overline{A}_3, 3A_2, 3m$
Trigonal	3m	Ditrigonal pyramidal	$1A_3, 3m$
division	32	Trigonal trapezohedral	$1A_3, 3A_2$
	$\overline{3}$	Rhombohedral	$i, 1\,\overline{A}_3$
	3	Trigonal pyramidal	$1A_3$
Orthorhombic	$2/m\,2/m\,2/m$	Rhombic dipyramidal	$i, 3A_2, 3m$
	mm2	Rhombic pyramidal	$1A_2, 2m$
	222	Rhombic disphenoidal	$3A_2$
Monoclinic	$2/m$	Prismatic	$i, 1A_2, 1m$
	m	Domatic	$1m$
	2	Sphenoidal	$1A_2$
Triclinic	$\overline{1}$	Pinacoidal	i
	1	Pedial	None

2

Study and Identification of Crystals

The students are supposed to study the crystal models in the practical class and identify a few of them in the examination. The following paragraphs deal with the description of different crystal forms. Attempt has also been made to acquaint the students with the identification procedures.

2.1 ISOMETRIC SYSTEM

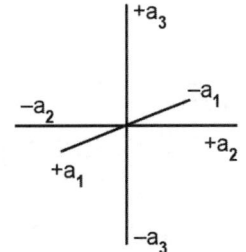

Fig. 2.1: Crystallographic axes of isometric system

Axial relationship: Crystals belonging to isometric or cubic system are referred to three crystallographic axes, which are all equal in length and are at right angles to each other (Fig. 2.1). Since the axes are mutually interchangeable, they are all designated by the letter 'a'. However, to distinguish them from each other and for proper orientation, they are designated as a_1, a_2 and a_3. In proper orientation, a_1 runs from front to back, positive at the observer side and negative at the back side of the crystal; a_2 runs from right to left of the observer, positive at the right-hand side and negative at the left-hand side; a_3 is vertical, positive at the top and negative at the bottom of the crystal. Axes a_1 and a_2 lie in the horizontal plane and a_3 is perpendicular to that plane. The orientation of the crystallographic axes and the method of their notation are shown in Fig. 2.1. The axial ratio (a: b: c) in case of isometric system is 1:1:1.

Five classes are grouped under this system. These are $4/m\,\overline{3}\,2/m$ (hexoctahedral), $\overline{4}\,3\,m$ (hextetrahedral), 432 (gyroidal), $2/m\,\overline{3}$ (diploidal) and 23 (tetartoidal). Characteristic symmetry elements of each class and different forms available in them are given below.

2.1.1 Class $4/m\,\overline{3}\,2/m$ (Hexoctahedral Class)

This is also known as the *normal class of the isometric system* as it shows maximum number of symmetry elements among all the classes of the isometric system. Some authors designate this class as *galena type* after the mineral galena that crystallizes in this class.

Symmetry elements: Crystals belonging to $4/m\,\overline{3}\,2/m$ class are characterised by three axes of 4-fold rotation, which are coincident with the crystallographic axes and perpendicular to the

8

cube faces at middle points. There are four axes of 3-fold rotoinversion, which emerge at the middle of each octant formed by intersection of crystallographic axes and join the opposite corners of the cube. In addition, there are six axes of 2-fold rotation, each of which bisects the angles between two crystallographic axes and join the middle points of opposite edges of the cube. There are nine mirror planes, three axial, each containing two of the crystallographic axes and six diagonal each of which bisects the angles between a pair of the axial planes. Centre of symmetry is present. This combination of symmetry elements is the highest symmetry possible in crystals.

Forms: Different forms belonging to this class are cube, octahedron, dodecahedron, tetrahexahedron, trisoctahedron, trapezohedron and hexoctahedron.

i. **Cube or Hexahedron $(001)_6$:** The cube (Fig. 2.2) is a solid bounded by six square faces each of which makes 90° angle with the adjacent face. Each face intersects one of the crystallographic axis and remains parallel with other two axes. The symbols of the faces towards the observer, to the right and at the top of the crystal are 100, 010 and 001 respectively. The symbols of the faces on the opposite sides of these faces and intersecting the a_1, a_2 and a_3 axes on the negative sides are $\bar{1}00, 0\bar{1}0$ and $00\bar{1}$ respectively. The form symbol is $(001)_6$.

ii. **Octahedron $(111)_8$:** The octahedron (Fig. 2.3) is a closed form bounded by eight equilateral triangular faces, each of which intersects all the three crystallographic axes at equal distances. The symbols of eight faces are $111, \bar{1}11, \bar{1}\bar{1}1, 1\bar{1}1, 11\bar{1}, \bar{1}1\bar{1}, \bar{1}\bar{1}\bar{1}$ and $1\bar{1}\bar{1}$. The form symbol is $(111)_8$.

iii. **Dodecahedron $(011)_{12}$:** The dodecahedron (Fig. 2.4) is a solid bounded by 12 rhomb-shaped faces, each of which intersects two of the crystallographic axes at equal distances and remains parallel with the third axis. Due to the rhombus shape of the faces, this form is commonly known as rhomb-dodecahedron. The symbols of twelve faces are $110, \bar{1}10, \bar{1}\bar{1}0, 1\bar{1}0, 101, \bar{1}01, \bar{1}0\bar{1}, 10\bar{1}, 011, 0\bar{1}1, 0\bar{1}\bar{1}$ and $01\bar{1}$. The form symbol is $(011)_{12}$.

iv. **Tetrahexahedron $(0kl)_{24}$:** The tetrahexahedron (Fig. 2.5) is a solid bounded by 24 isosceles triangular faces, each of which intersects two of the crystallographic axes at unequal lengths and remains parallel with the third axis. Hence the form symbol is $(0kl)_{24}$. Four faces of the tetrahexahedron occupy one face of hexahedron (cube) making a small pyramid, hence the name. There can be many tetrahexahedrons depending on the values of k and l. The most common form is $(012)_{24}$, though forms like $(013)_{24}, (014)_{24}, (023)_{24}$, etc. also occur.

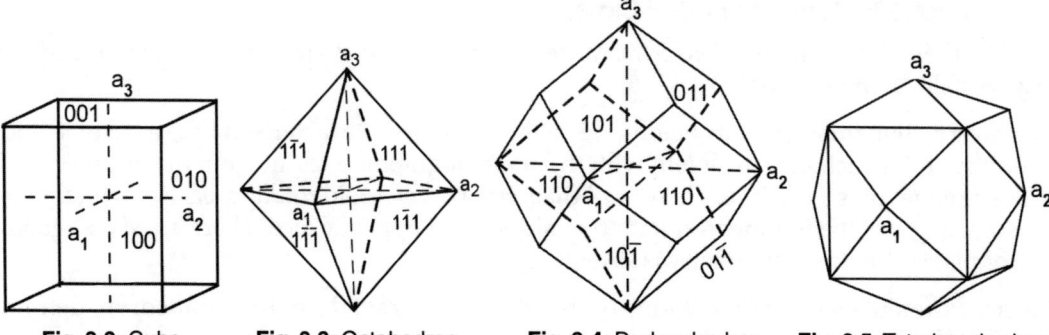

Fig. 2.2: Cube   **Fig. 2.3:** Octahedron   **Fig. 2.4:** Dodecahedron   **Fig. 2.5:** Tetrahexahedron

v. Trisoctahedron (hll)$_{24}$: The trisoctahedron (Fig. 2.6) is a solid having 24 isosceles triangular faces, each of which intersects two of the crystallographic axes at equal lengths and the third axis at a greater distance. Thus, the form symbol is (hll)$_{24}$. The most common form is (122)$_{24}$ though forms like (133)$_{24}$ and (233)$_{24}$ do occur. It has the appearance of being formed by a three-faced pyramid grown on each face of the octahedron. It is also known as trigonal trisoctahedron due to the fact that each face has three edges and three of the faces occupy one face of the octahedron.

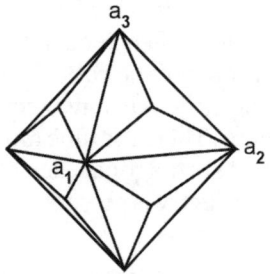

Fig. 2.6: Trisoctahedron

vi. Trapezohedron (hhl)$_{24}$: The trapezohedron (Fig. 2.7) is a solid bounded by 24 trapezoid faces, each of which intersects two of the crystallographic axes at equal lengths and the third axis at a shorter distance. Thus, the form symbol is (hhl)$_{24}$. Depending on the values of h and l, there are various trapezohedrons like (112)$_{24}$, (113)$_{24}$, (114)$_{24}$, (223)$_{24}$, but the form with symbol (112)$_{24}$ is most common. Since three faces of this form, each having four edges, occupy one face of the octahedron, this form is also called tetragonal trisoctahedron.

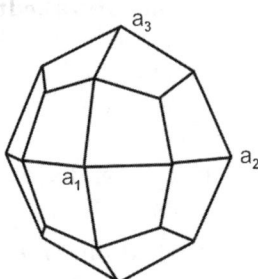

Fig. 2.7: Trapezohedron

vii. Hexoctahedron (hkl)$_{48}$: The hexoctahedron (Fig. 2.8) is a solid having 48 scalene triangular faces, each of which intersects all the three crystallographic axes at different lengths. Thus, the general form symbol is (hkl)$_{48}$. Depending on the values of h, k and l, different hexoctahedrons exist, but the most common form is (123)$_{48}$. From Fig. 2.8 it is apparent that six of the faces occur on one face of the octahedron, hence the name.

Minerals crystallizing in this class are diamond, native platinum, gold, silver and copper, magnetite, galena, fluorite, garnet, cerargyrite, cuprite, halite, sylvite, uraninite, leucite, etc.

Various forms of hexoctahedral class also occur in combinations with each other. In such cases, the shapes of the faces of the simple forms get obliterated but the crystallographic character like intercept relationship of each form remains intact.

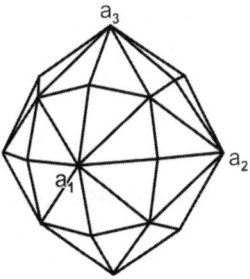

Fig. 2.8: Hexoctahedron

2.1.2 Class $\bar{4}$3m (Hextetrahedral Class)

Some authors designate this class *tetrahedrite type* after the characteristic mineral tetrahedrite that crystallizes in this class.

Symmetry elements: Crystals belonging to this class are characterised by 3 axes of 4-fold rotoinversion, which are coincident with the crystallographic axes and join the midpoints of opposite edges of tetrahedron. There are 4 diagonal axes of 3-fold rotation, each of which joins the solid angle of the tetrahedron with the midpoint of opposite face. There are six diagonal mirror planes. Centre of symmetry is absent.

Forms: Different characteristic forms of this class are tetrahedron, tristetrahedron, deltoid dodecahedron and hextetrahedron.

i. Tetrahedron (111)₄: Tetrahedron is a solid bounded by four equilateral triangular faces, each of which intersects all the three crystallographic axes at equal lengths. Therefore, the form symbol is $(111)_4$. The tetrahedron can be considered as derived from the octahedron of the $4/m\,\overline{3}\,2/m$ class by omission of the alternate faces and the extension of the others. Thus, corresponding to one octahedron, there are two tetrahedrons, one designated as positive (Fig. 2.9) with faces $111,\ \overline{1}\,\overline{1}1,\ 1\overline{1}\,\overline{1}$ and $\overline{1}1\overline{1}$ and the other designated as negative (Fig. 2.10) with faces $\overline{1}11, 1\overline{1}1, 11\overline{1}$ and $\overline{1}\,\overline{1}\,\overline{1}$. The form symbols of positive and negative tetrahedrons are $(111)_4$ and $(1\overline{1}1)_4$ respectively. These two tetrahedrons are geometrically alike and one can be converted to the other by rotation of 90° about the vertical crystallographic axis. In case of positive tetrahedron the top edge runs from left to right of the observer, whereas in case of negative tetrahedron the top edge is oriented front to back of the crystal.

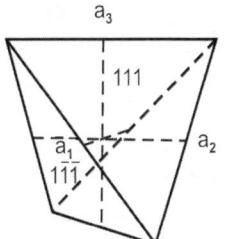

Fig. 2.9: Positive tetrahedron

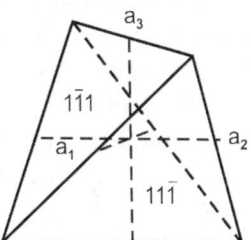

Fig. 2.10: Negative tetrahedron

ii. Tristetrahedron (hhl)₁₂: The tristetrahedron is a solid bounded by 12 isosceles triangular faces, each of which intersects two crystallographic axes at equal lengths and the third at a shorter distance. It has the appearance of having a low three-faced pyramid raised on each face of the tetrahedron. Corresponding to the trapezohedron of the $4/m\,\overline{3}\,2/m$ class, there are two tristetrahedrons designated as positive (Fig. 2.11) with form symbol $(hhl)_{12}$ and negative with the form symbol $(h\overline{h}l)_{12}$. Both the forms are geometrically alike and the positive form can be made negative by rotation of 90° about the vertical crystallographic axis and vice versa.

iii. Deltoid dodecahedron (hll)₁₂: The deltoid dodecahedron is a solid bounded by 12 trapezoid faces, each of which intersects two crystallographic axes at equal lengths and the third at a greater distance. Corresponding to the trisoctahedron of the $4/m\,\overline{3}\,2/m$ class, there are two deltoid dodecahedrons designated as positive (Fig. 2.12) with form symbol $(hll)_{12}$ and negative with the form symbol $(h\overline{h}l)_{12}$. Both the forms are geometrically alike and the positive form can be made negative by rotation of 90° about the vertical crystallographic axis and vice versa.

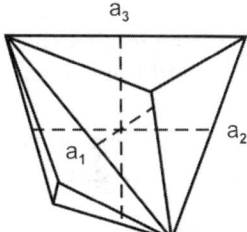

Fig. 2.11: Positive tristetrahedron

iv. Hextetrahedron (hkl)₂₄: The hextetrahedron is a solid bounded by 24 scalene triangular faces, each of which intersects all the three crystallographic axes at unequal lengths. Six of the faces occupy one face of the tetrahedron, hence the name. Corresponding to the hexoctahedron of the $4/m\,\overline{3}\,2/m$ class, there are two hextetrahedrons designated as positive (Fig. 2.13) with form symbol $(hkl)_{24}$ and negative with form symbol $(h\overline{k}l)_{24}$.

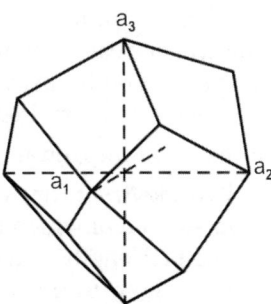

Fig. 2.12: Positive deltoid dodecahedron

Other forms occurring in the hextetrahedral class are cube $(001)_6$, dodecahedron $(011)_{12}$ and tetrahexahedron $(0kl)_{24}$. Though geometrically similar to the corresponding forms of $4/m\bar{3}2/m$ class, these forms have lower structural symmetry expressed by the development of striations, etch-marks or by pyroelectricity property. A striated cube is shown in Fig. 2.14.

Minerals crystallizing in this class are the members of tetrahedrite—tennantite series, sphalerite and boracite.

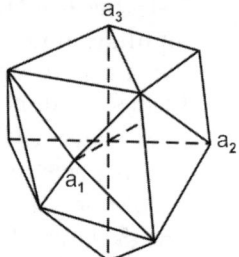

Fig. 2.13: Positive hextetrahedron

2.1.3 Class 432 (Gyroidal Class)

Symmetry elements: The gyroidal class has 3 axes of 4-fold rotational symmetry coincident with the crystallographic axes. In addition, there are 4 axes of 3-fold and 6 axes of 2-fold rotational symmetry. The class is devoid of plane and centre of symmetry.

Forms: Gyroid is the only general form of this class. It is a solid bounded by 24 pentagonal faces, each of which intersects all the crystallographic axes at unequal lengths. Corresponding to the hexoctahedron of the $4/m\bar{3}2/m$ class, there are two gyroids designated as right-handed (Fig. 2.15) with form symbol $(hkl)_{24}$ and left-handed (Fig. 2.16) with the form symbol $(khl)_{24}$. These two forms display enantiomorphic relationship.

Fig. 2.14: Striated cube

Other forms occurring in the gyroidal class are cube $(001)_6$, octahedron $(111)_8$, dodecahedron $(011)_{12}$, tetrahexahedron $(0kl)_{24}$, trisoctahedron $(hll)_{24}$ and trapezohedron $(hhl)_{24}$. Though geometrically similar with the corresponding forms of $4/m\bar{3}2/m$ class, these forms have lower structural symmetry expressed by the development of striations, etch-marks etc. There is no representative mineral in the gyroidal class.

2.1.4 Class 2/m$\bar{3}$ (Diploidal Class)

This class is also known *pyrite type* after the characteristic mineral pyrite that crystallizes in this class.

Symmetry elements: The three crystallographic axes are the axes of 2-fold rotation. In addition, there are four diagonal axes of 3-fold rotoinversion each of which emerges in the middle of the octant formed by crystallographic axes. The planes of symmetry are three in number and axial in position. Centre of symmetry is present.

Forms: Pyritohedron and diploid are the two typical forms of this class.

  i. Pyritohedron: The pyritohedron is a solid bounded by 12 pentagonal faces each of which intersects two of the crystallographic axes at unequal lengths and remains parallel with the third axis. Out of the five edges, four are equal in length and the fifth one is longer. Corresponding to the tetrahexahedron of the $4/m\bar{3}2/m$ class there are two pyritohedrons, positive $(h0l)_{12}$ (Fig. 2.17) and negative $(0kl)_{12}$

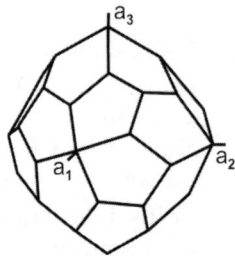

Fig. 2.15: Right handed gyroid

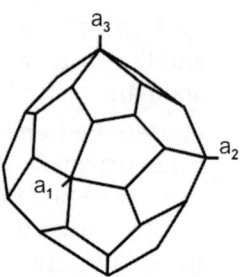

Fig. 2.16: Left handed gyroid

(Fig. 2.18) produced from the tetrahexahedron by the development of alternate faces. In positive pyritohedron the longer edge towards the observer is vertical, whereas in negative pyritohedron the longer edge towards the observer is horizontal. A rotation of 90° about a crystallographic axis brings the positive pyritohedron into the negative position and vice versa.

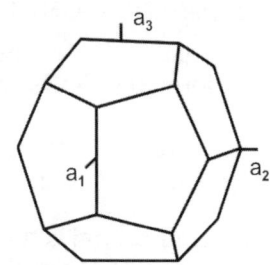

Fig. 2.17: Positive pyritohedron

ii. **Diploid:** The diploid is a solid bounded by 24 trapezoid faces, each of which intersects all the three crystallographic axes at unequal lengths. The faces occur in pairs and the diploid may be thought of as originated from the pyritohedron by development of a pair of faces on each face of the pyritohedron. Corresponding to the hexoctahedron of the $4/m\,\overline{3}\,2/m$ class, there are two diploids, positive $(hkl)_{24}$ (Fig. 2.19) and negative $(khl)_{24}$ produced from the hexoctahedron by development of alternate faces.

Other forms occurring in the diploidal class are cube $(001)_6$, octahedron $(111)_8$, dodecahedron $(011)_{12}$, trisoctahedron $(hll)_{24}$ and trapezohedron $(hhl)_{24}$. Though geometrically similar to the corresponding forms of hexoctahedral $(4/m\,\overline{3}\,2/m)$ class, these forms have lower structural symmetry expressed by the development of striations, etch marks, etc.

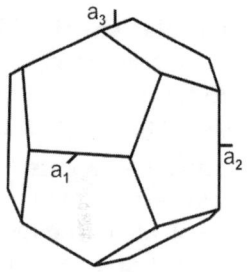

Fig. 2.18: Negative pyritohedron

The chief mineral crystallizing in this class is pyrite in addition to rare minerals like smaltite, chloanthite, sperrylite, etc.

2.1.5 Class 23 (Tetartoidal Class)

Symmetry elements: The three crystallographic axes are the axes of two-fold rotation and the four diagonal axes are the axes of 3-fold rotational symmetry. Plane of symmetry and centre of symmetry are absent.

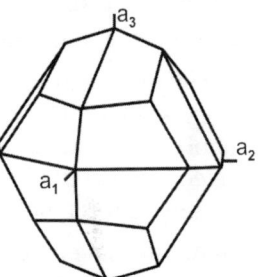

Fig. 2.19: Positive diploid

Forms: Tetartoid is the only general form of this class. It is a solid bounded by 12 pentagonal faces, each of which intersects all the crystallographic axes at unequal lengths. Corresponding to the hexoctahedron of the $4/m\,\overline{3}\,2/m$ class, there are four tetartoids which are designated as positive right $(hkl)_{12}$, positive left $(khl)_{12}$, negative right $(k\,\overline{h}\,l)_{12}$ and negative left $(h\,\overline{k}\,l)_{12}$. They comprise two enantiomorphic pairs, positive right and left, and negative right and left. The positive left and right tetartoids are shown in Figs 2.20 and 2.21 respectively.

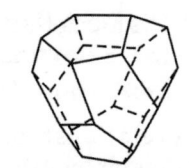

Fig. 2.20: Positive left tetartoid

Other forms present in this class are cube, dodecahedron, pyritohedron, tetrahedron and deltoid dodecahedron. These forms have lower structural symmetry expressed by the development of striations, etch marks, etc.

Cobaltite and ullmanite are the most common minerals crystalising in this class in addition to artificial crystals of barium nitrate, strontium nitrate, sodium chloride, etc.

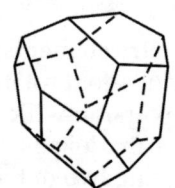

Fig. 2.21: Positive right tetartoid

2.2 IDENTIFICATION OF ISOMETRIC CRYSTAL FORMS

Nine combination forms are shown in Figs 2.22–2.30. These are to be identified with respect to their form, class and system. It is to be noted that, the system and class to which a form belongs are worked out from axial relationship and symmetry elements of the concerned form respectively. The types of the forms present are determined from the number of faces and intercept relationships.

2.2.1 The Model is a Combination of Two Sets of Faces, 6 Square-shaped (A) and 12 Hexagonal (B), Each of which Represents a Form (Fig. 2.22)

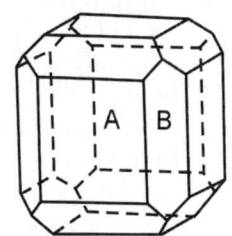

Fig. 2.22: Isometric combination form

Axial relationship: The combination form can be referred to three crystallographic axes, which are equal in length and perpendicular to each other. These axes join the midpoints of opposite square shaped faces. Thus, the form represented by Fig. 2.22 belongs to isometric system.

Symmetry elements: Examination of the form indicates the presence of 3 axes of 4-fold rotation which are coincident with crystallographic axes, 4 axes of 3-fold rotoinversion which join the opposite solid angles where three hexagonal faces meet and 6 axes of 2-fold rotation that join the midpoints of opposite hexagonal faces. There are three axial planes, each containing two of the crystallographic axes and six diagonal planes that divide the angles between axial planes. Centre of symmetry is also present. This set of symmetry elements is the characteristic of the hexoctrahedral ($4/m\,\overline{3}\,2/m$) class.

Forms: There are two sets of faces in the model. The square-shaped faces are six in number, each intersects one of the crystallographic axis remaining parallel with other two. Thus, the form symbol is $(001)_6$ and it is cube. The hexagonal faces are 12 in number, when produced each face intersects two of the crystallographic axes at equal lengths remaining parallel with the third. Thus, the form symbol is $(011)_{12}$ and it is dodecahedron.

Conclusion: The model shown in Fig. 2.22 is a combination of cube and dodecahedron of hexoctrahedral ($4/m\,\overline{3}\,2/m$) class of isometric system.

2.2.2 The Model is a Combination of three sets of Faces, 6 Octagonal (A), 8 Hexagonal (B), and 12 Rectangular (C) Each of which Represents a Form (Fig. 2.23)

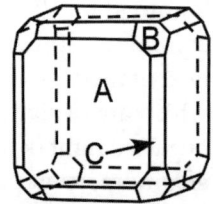

Fig. 2.23: Isometric combination form

Axial relationship: The combination form can be referred to three crystallographic axes, which are equal in length and perpendicular to each other. These axes join the midpoints of opposite octagonal faces. Thus, the form represented by Fig. 2.23 belongs to isometric system.

Symmetry elements: Examination of the form indicates the presence of 3 axes of 4-fold rotation which are coincident with crystallographic axes, 4 axes of 3-fold rotoinversion which join the midpoints of opposite hexagonal faces and 6 axes of 2-fold rotation that join the midpoints of opposite rectangular faces. There are three axial planes, each containing two of the crystallographic axes and six diagonal planes that divide the angles between axial planes. Centre of symmetry is also present. This set of symmetry elements is the characteristic of the hexoctrahedral class ($4/m\,\overline{3}\,2/m$).

Forms: There are three sets of faces in the model. The octagonal faces are six in number, each intersects one of the crystallographic axis remaining parallel with other two. Thus, the form symbol is $(001)_6$ and it is cube. The hexagonal faces are eight in number, each intersects all the three crystallographic axes at equal lengths when extended. Thus, the form symbol is $(111)_8$ and it is octahedron. The rectangular faces are 12 in number, when produced, each face intersects two of the crystallographic axes at equal lengths remaining parallel with the third. Thus, the form symbol is $(011)_{12}$ and it is dodecahedron.

Conclusion: The model shown in Fig. 2.23 is a combination of cube, octahedron and dodecahedron of hexoctrahedral class $(4/m\,\overline{3}\,2/m)$ of isometric system.

2.2.3 By the Considerations Stated for Identification of Model shown in Fig. 2.23, it can be Proved that the Model shown in Fig. 2.24 is a Combination of Cube, Octahedron and Dodecahedron of Hexoctrahedral Class (4/m 3̄ 2/m) of Isometric System (Fig. 2.24)

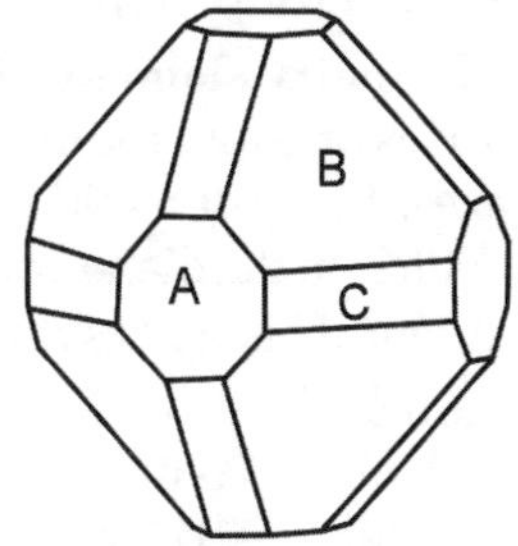

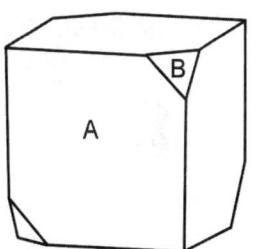

Fig. 2.24: Isometric combination form

A close look at the above models shown in Figs 2.22–2.24 indicates that the shape of different faces in combination forms vary appreciably depending on their development. The square faces of the cube in Fig. 2.22, changed to octagonal in Fig. 2.23 due to the addition of octahedron. The relative size of these octagonal cube faces in Fig. 2.23 reduced in size in Fig. 2.24 due to expansion of octahedron faces. Similarly, the hexagonal faces of dodecahedron in Fig. 2.22 changed to rectangular shape in Figs 2.23 and 2.24 due to addition of octahedron. The difference in the orientation of these square faces in Figs 2.23 and 2.24 is attributed to the relative enlargement of cube and octahedron faces.

2.2.4 The Form is a Combination of Two Sets of Faces, Hexagonal (A) and Triangular (B) (Fig. 2.25)

Axial relationship: The combination form can be referred to three crystallographic axes, which are equal in length and perpendicular to each other. These axes join the midpoints of opposite hexagonal faces. Thus, the form represented by Fig. 2.25 belongs to isometric system.

Symmetry elements: Examination of the form indicates the presence of 3 axes of 4-fold rotoinversion, which are coincident with crystallographic axes, 4 axes of 3-fold rotation, which join the midpoints of triangular faces with the solid angles on the diametrically opposite side. There are six diagonal planes. Centre of symmetry is absent. This set of symmetry elements is the characteristic of the class $\overline{4}\,3m$ (hextetrahedral).

Fig. 2.25: Isometric combination form

Forms: There are two sets of faces in the model. The hexagonal faces are six in number, each intersects one of the crystallographic axis remaining parallel with other two. Thus, the form symbol is $(001)_6$ and it is cube. The triangular faces are four in number, each intersects all the three crystallographic axes at equal lengths when extended. One of the faces has symbol of 111. Thus, the form symbol is $(111)_4$ and it is the positive tetrahedron.

Conclusion: The model is a combination of cube and positive tetrahedron of hextetrahedral class $(\overline{4}\,3m)$ of isometric system. However, if the model is rotated by 90° about the vertical axis, it will represent the combination of cube and negative tetrahedron.

2.2.5 The Form is a Combination of three Sets of Faces, Bigger Hexagonal (A), Smaller Hexagonal (B) and Pentagonal (C) (Fig. 2.26)

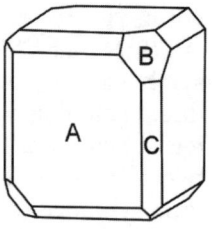

Fig. 2.26: Isometric combination form

Axial relationship: The combination form can be referred to three crystallographic axes, which are equal in length and perpendicular to each other. These axes join the midpoints of opposite bigger hexagonal faces. Thus, the form represented by Fig. 2.26 belongs to isometric system.

Symmetry elements: Examination of the form indicates the presence of 3 axes of 4-fold rotoinversion, which are coincident with crystallographic axes, 4 axes of 3-fold rotation that join the midpoints of smaller hexagonal faces with the solid angles on the diametrically opposite side. There are six diagonal planes. Centre of symmetry is absent. This set of symmetry elements is the characteristic of the class $\bar{4}$3m (hextetrahedral).

Forms: There are three sets of faces in the model. The bigger hexagonal faces are six in number, each intersects one of the crystallographic axis remaining parallel with other two. Thus, the form symbol is $(001)_6$ and it is cube. The smaller hexagonal faces are four in number, each intersects all the three crystallographic axes at equal lengths when extended. One of the faces has symbol of 111. The form symbol is $(111)_4$ and it is positive tetrahedron. The pentagonal faces are twelve in number, each intersects two of the crystallographic axes at equal lengths remaining parallel with the third axis when extended. The form symbol is $(011)_{12}$. It is dodecahedron.

Conclusion: The model shown in Fig. 2.26 is a combination of cube, dodecahedron and positive tetrahedron of class $\bar{4}$3m (hextetrahedral) of isometric system. However, if the model is rotated by 90° about the vertical axis, it will represent the combination of cube, dodecahedron and negative tetrahedron.

2.2.6 The Crystal Model shown in Fig. 2.27 is a Combination of Two Sets of Faces, Six Square (A) and Twenty-four Pentagonal (B) (Fig. 2.27)

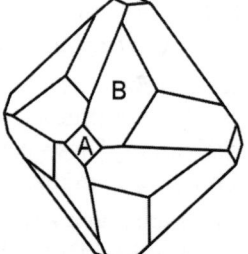

Fig. 2.27: Isometric combination form

Axial relationship: The crystal model can be referred to three equal and mutually perpendicular axes, which join the midpoints of opposite square faces and thus, belong to the isometric system.

Symmetry elements: Symmetry elements include 3 axes of 4-fold, 4 axes of 3-fold and 6 axes of 2-fold rotational symmetry without plane and centre that characterise the gyroidal class.

Forms: The smaller rectangular faces (A) are six in number, each intersects one of the crystallographic axis remaining parallel with other two. The form symbol is $(001)_6$ and it is cube. The pentagonal faces (B) are 24 in number, each intersects all the three crystallographic axes at unequal lengths and represent right-handed gyroid.

Conclusion: The model shown in Fig. 2.27 is a combination of cube and right-handed gyroid belonging to the gyroidal class of the isometric system.

2.2.7 The Form is a Combination of Two Sets of Faces, Rectangular (A) and Hexagonal (B) (Fig. 2.28)

Axial relationship: The combination form can be referred to three crystallographic axes, which are equal in length and perpendicular to each other. These axes join the midpoints of opposite rectangular faces. Thus, the form represented by Fig. 2.28 belongs to the isometric system.

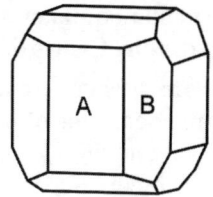

Fig. 2.28: Isometric combination form

Symmetry elements: The three crystallographic axes are the axes of 2-fold rotation. In addition, there are four diagonal axes of 3-fold rotoinversion each of which emerges at the solid angle formed by the intersection of three hexagonal faces. The planes of symmetry are three in number and axial in position. Centre of symmetry is present. Thus, the form belongs to $(2/m\,\overline{3})$ diploidal class.

Forms: There are two sets of faces in the model. The rectangular faces are six in number, each intersects one of the crystallographic axis remaining parallel with other two. The form symbol is $(001)_6$ and it is the cube. The hexagonal faces are twelve in number, each intersects two of the crystallographic axes at unequal lengths remaining parallel with the third. The form symbol is $(h0l)_{12}$. It is the positive pyritohedron.

Conclusion: The model shown in Fig. 2.28 is a combination of cube and positive pyritohedron of diploidal class $(2/m\,\overline{3})$ of isometric system.

2.2.8 The Form is a Combination of Two Sets of Faces, Six 12-Sided (A) and Twenty-four Quadrangular (B) (Fig. 2.29)

Axial relationship: The combination form can be referred to three crystallographic axes, which are equal in length and perpendicular to each other. These axes join the midpoints of opposite 12-sided faces. Thus, the form represented by Fig. 2.29 belongs to isometric system.

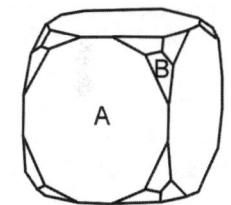

Fig. 2.29: Isometric combination form

Symmetry elements: The three crystallographic axes are the axes of 2-fold rotation. In addition there are four diagonal axes of 3-fold rotoinversion each of which emerges at the solid angle formed by the intersection of three quadrangular faces. The planes of symmetry are three in number and axial in position. Centre of symmetry is present. Thus, the form belongs to $(2/m\,\overline{3})$ diploidal class.

Forms: There are two sets of faces in the model. The 12-sided faces are six in number, each intersects one of the crystallographic axis remaining parallel with other two. The form symbol is $(001)_6$ and it is the cube. The quadrangular faces are 24 in number, each intersects all the crystallographic axes at unequal lengths. The form symbol is $(hkl)_{24}$. It is the positive diploid.

Conclusion: The model shown in Fig. 2.29 is a combination of cube and positive diploid of class $2/m\,\overline{3}$ (diploidal) of isometric system.

2.2.9 The Form is a Combination of Four Sets of Faces, Hexagonal (A), Smaller Equilateral Triangular (B), Bigger Isosceles Triangular (C) and Heptagonal (D) (Fig. 2.30)

Axial relationship: The combination form can be referred to three crystallographic axes, which are equal in length and perpendicular to each other. These axes join the midpoints of opposite hexagonal faces. Thus, the form represented by Fig. 2.30 belongs to isometric system.

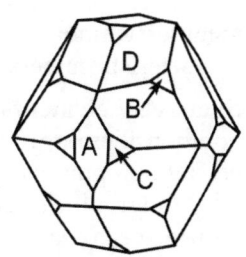

Fig. 2.30: Isometric combination form

Symmetry elements: The three crystallographic axes are the axes of 2-fold rotation. In addition, there are four diagonal axes of 3-fold rotoinversion each of which emerges at the midpoint of opposite smaller equilateral triangular faces. The planes of symmetry are three in number and axial in position. Centre of symmetry is present. Thus, the form belongs to diploidal class.

Forms: There are four sets of faces in the model. The hexagonal faces are six in number, each intersects one of the crystallographic axis remaining parallel with other two. Thus, the form symbol is $(001)_6$ and it is the cube. The smaller equilateral triangular faces are 8 in number, each intersects all the crystallographic axes at equal lengths. The form symbol is $(111)_8$. It is octahedron. The bigger isosceles triangular faces are 12 in number, each intersects two of the crystallographic axes at unequal lengths remaining parallel with the third. The form symbol is $(h0l)_{12}$ and it is positive pyritohedron. The heptagonal faces are 24 in number, each intersects all the crystallographic axes at unequal lengths. The form symbol is $(hkl)_{24}$ and it is the positive diploid.

Conclusion: The model shown in Fig. 2.30 is a combination of cube, octahedron, positive pyritohedron and positive diploid of diploidal class ($2/m\,\overline{3}$) of isometric system.

2.3 TETRAGONAL SYSTEM

Axial relationship: Crystals belonging to tetragonal system are referred to three crystallographic axes, which are perpendicular to each other (Fig. 2.31). Out of the three axes, two are horizontal, lie in the horizontal plane, equal in length and thus, are mutually interchangeable. These are designated as a_1 and a_2. Axis a_1 runs from front to back, positive at observer side and negative in the back; a_2 axis runs from right to left, positive at right hand side and negative at left hand side. The third axis is vertical; perpendicular to the plane containing a_1 and a_2 axes and is different in length, either shorter as in case of zircon or longer as in case of octahedrite. It is designated as 'c'. The axial ratio (a:c) in case of zircon and octahedrite are 1:0.901 and 1:1.777 respectively. The axial relationship of the crystals belonging to tetragonal system is shown in Fig. 2.31.

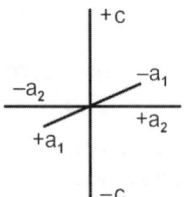

Fig. 2.31:
Crystallographic axes of tetragonal system

Seven classes are grouped under this system. These are $4/m\,2/m\,2/m$ (ditetragonal dipyramidal), $\overline{4}\,2m$ (tetragonal scalenohedral), $4\,mm$ (ditetragonal pyramidal), 422 (tetragonal trapezohedral), $4/m$ (tetragonal dipyramidal), $\overline{4}$ (tetragonal disphenoidal) and 4 (tetragonal pyramidal). Characteristic symmetry elements of each class and different forms available in them are given below.

2.3.1 Class 4/m 2/m 2/m (Ditetragonal Dipyramidal Class)

This class is also known as the *normal class of tetragonal system* as it possesses maximum symmetry among all the classes belonging to tetragonal system and *zircon type* after the characteristic mineral zircon that crystallizes in this class.

Symmetry elements: The vertical crystallographic axis 'c' is the axis of 4-fold rotation. In addition, there are four axes of 2-fold symmetry, two of which are coincident with the horizontal crystallographic axes and other two are diagonal bisecting the angles between the horizontal crystallographic axes. There are five mirror planes, three are axial (one horizontal and two vertical) each containing two of the crystallographic axes and other two are vertically diagonal bisecting the angles between the vertical axial planes. Centre of symmetry is present.

Forms: Different forms belonging to this class are basal pinacoid, three types of prisms and corresponding three types of dipyramids.

i. **Basal pinacoid (001)$_2$:** The basal pinacoid (Fig. 2.32) is an open form composed of two horizontal faces, which are parallel with the horizontal axes and intersect the vertical crystallographic axis at equal lengths. The form symbol is (001)$_2$ and the two faces are 001 and 00$\bar{1}$. As it forms the base of the crystal, it is known as basal pinacoid.

ii. **Tetragonal prism of first order (110)$_4$:** The tetragonal prism of first order (Fig. 2.33) is an open form that consists of four rectangular faces, each of which intersects the horizontal crystallographic axes at equal lengths and remains parallel with the vertical crystallographic axis. The form symbol is (110)$_4$ and the four faces are 110, $\bar{1}$10, $\bar{1}$ $\bar{1}$0 and 1$\bar{1}$0.

iii. **Tetragonal prism of second order (100)$_4$:** The tetragonal prism of second order (Fig. 2.34) is an open form that consists of four rectangular faces, each of which intersects one horizontal crystallographic axis and remains parallel with the other two crystallographic axes. The form symbol is (100)$_4$ and the four faces are 100, 010, $\bar{1}$00 and 0$\bar{1}$0.

iv. **Ditetragonal prism (hk0)$_8$:** The ditetragonal prism (Fig. 2.35) is an open form that consists of eight rectangular faces, each of which intersects both the horizontal crystallographic axes at different lengths. The form symbol is (hk0)$_8$. This prism can be thought of as derived from the tetragonal prism of first order by development of two faces on each face of the tetragonal prism of first order. Depending upon the different intercepts made on the horizontal crystallographic axes a number of ditetragonal prisms exist but the most common one is (120)$_8$.

The relationship between three tetragonal prisms is shown in Fig. 2.36, which is a cross section in the horizontal plane of symmetry. Both the prisms of first and second order are geometrically alike and interchangeable with a rotation by 45° about the vertical crystallographic axis. In case of prism of first order, the horizontal axes emerge at the

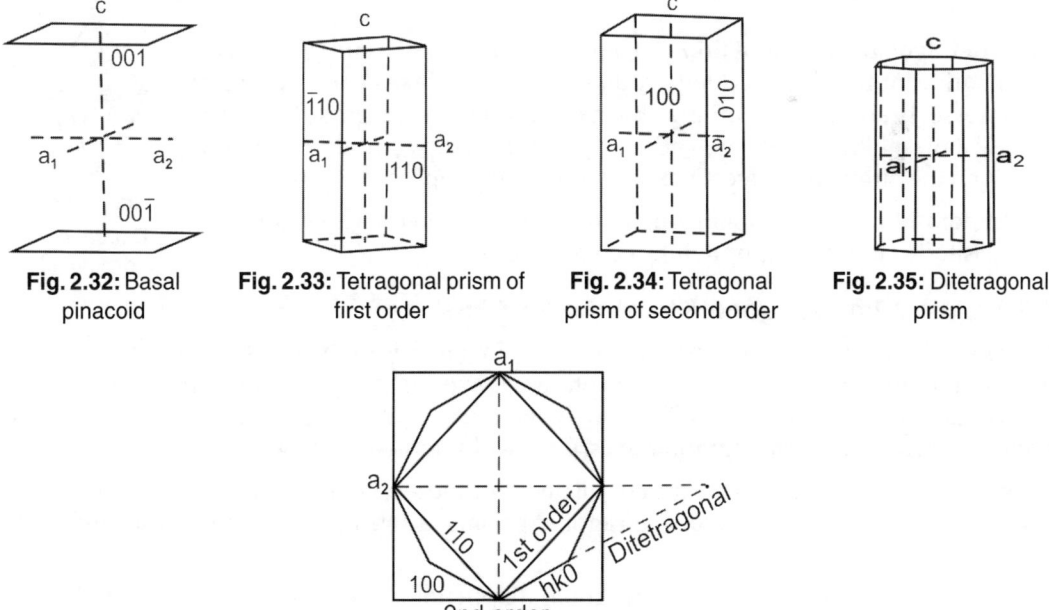

Fig. 2.32: Basal pinacoid   **Fig. 2.33:** Tetragonal prism of first order   **Fig. 2.34:** Tetragonal prism of second order   **Fig. 2.35:** Ditetragonal prism

Fig. 2.36: Relationship between three types of tetragonal prisms

midpoints of opposite vertical edges while in case of prism of second order, the horizontal axes are perpendicular to the prism faces at their midpoints.

v. Tetragonal dipyramid of first order (hhl)$_8$: The tetragonal dipyramid of first order (Fig. 2.37) is a closed form having eight isosceles triangular faces, each of which intersects both the horizontal crystallographic axes at equal lengths and the vertical crystallographic axis at a different length. The form symbol is (hhl)$_8$. If a face of this dipyramid cuts the horizontal and vertical crystallographic axes in the same ratio as the axial ratio, the form is called *unit* or *fundamental* form and the symbol becomes (111)$_8$. Though the unit dipyramid is most common, dipyramids with symbols (221)$_8$, (331)$_8$, (112)$_8$, (113)$_8$, etc also exist.

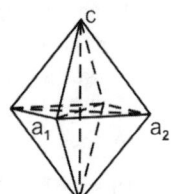

Fig. 2.37: Tetragonal dipyramid of first order

vi. Tetragonal dipyramid of second order (0kl)$_8$: The tetragonal dipyramid of second order (Fig. 2.38) is a closed form having eight isosceles triangular faces, each of which intersects the vertical crystallographic axis and one of the horizontal crystallographic axis at different lengths remaining parallel with the other horizontal crystallographic axis. The form symbol is (0kl)$_8$. The unit dipyramid (011)$_8$ is most common, but dipyramids with symbols (021)$_8$, (031)$_8$, (012)$_8$, (013)$_8$, etc also exist.

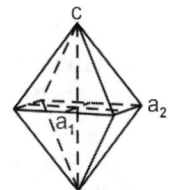

Fig. 2.38: Tetragonal dipyramid of second order

As a general rule, if only one dipyramid is present, it is set as first order. However, if two dipyramids of different orders are present, the dominant one is usually set as first order. In the orientation of a crystal, the prisms are subordinate to the dipyramids. Thus, an important prism with bigger faces may be relegated to second order by the presence of a small dipyramid.

vii. Ditetragonal dipyramid (hkl)$_{16}$: The ditetragonal dipyramid (Fig. 2.39) is a solid bounded by sixteen scalene triangular faces each of which intersects all the crystallographic axes at unequal lengths. The form symbol is (hkl)$_{16}$. There are various ditetragonal dipyramids depending upon the different intersections on the crystallographic axes.

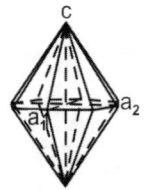

Fig. 2.39: Ditetragonal dipyramid

Important minerals crystallizing in this class are zircon, rutile, anatase, cassiterite, idocrase, apophyllite, octahedrite, etc.

2.3.2 Class $\bar{4}$ 2m (Tetragonal Scalenohedral Class)

Symmetry elements: The vertical crystallographic axis 'c' is the axis of 4-fold rotoinversion. In addition, there are two axes of 2-fold rotational symmetry, which are coincident with the horizontal crystallographic axes. There are two vertical mirror planes, which bisect the angles between horizontal crystallographic axes. Centre of symmetry is absent.

Forms: There are two special forms, viz. disphenoid and scalenohedron in addition to the basal pinacoid, first- and second-order tetragonal prisms, ditetragonal prism and second-order tetragonal dipyramid.

i. Tetragonal disphenoid: The tetragonal disphenoid is a closed form bounded by four isosceles triangular faces each of which intersects both the horizontal crystallographic axes at equal lengths and the vertical crystallographic axis at a different length. The form symbol is $(hhl)_4$. Corresponding to the first-order tetragonal dipyramid there are two disphenoids, positive (Fig. 2.40) with form symbol $(hhl)_4$ and negative (Fig. 2.41) with form symbol $(h\bar{h}l)_4$.

ii. **Tetragonal scalenohedron:** The tetragonal scalenohedron is a solid bounded by eight scalene triangular faces each of which intersects all the three crystallographic axes at unequal lengths. It may be considered as derived from the ditetragonal dipyramid by taking alternate pairs of faces. There are two tetragonal scalenohedra, positive with form symbol $(hkl)_8$ and negative with form symbol $(h\bar{k}l)_8$. The positive tetragonal scalenohedron is shown in Fig. 2.42.

Chalcopyrite and stannite are the common minerals crystallising in this class.

2.3.3 Class 4 mm (Ditetragonal Pyramidal Class)

Symmetry elements: This class is characterised by the presence of one axis of 4-fold rotation coincident with the vertical crystallographic axis. In addition, there are four vertical planes, two axial and two diagonal.

Forms: The lack of a horizontal plane of symmetry gives rise to different forms at the top and bottom of crystals, which are hemimorphic forms. These forms are pedion and tetragonal pyramids.

i. **Pedion:** It is an open form having only one face that intersects the c-crystallographic axis. Corresponding to the basal pinacoid of the $4/m\,2/m\,2/m$ class, there are two pedions, upper $(001)_1$ shown in Fig. 2.43 and lower $(00\bar{1})_1$.

ii. **Tetragonal pyramid of first order:** It is an open form having four isosceles triangular faces each of which intersects both the horizontal crystallographic axes at equal lengths and the vertical crystallographic axis at different length. Corresponding to the tetragonal dipyramid of first order of the $4/m\,2/m\,2/m$ class, there are two pyramids, upper $(hhl)_4$ shown in Fig. 2.44 and lower $(hh\bar{1})_4$.

iii. **Tetragonal pyramid of second order:** It is an open form having four isosceles triangular faces each of which intersects one of the horizontal crystallographic axis and the vertical crystallographic axis at different lengths. Corresponding to the tetragonal dipyramid of second order of the $4/m\,2/m\,2/m$ class, there are two pyramids, upper $(h0l)_4$ shown in Fig. 2.45 and lower $(h0\bar{1})_4$.

iv. **Ditetragonal pyramid:** It is an open form having eight scalene triangular faces each of which intersects all the three crystallographic axes at different lengths. Corresponding to the ditetragonal dipyramid of the $4/m\,2/m\,2/m$ class, there are two pyramids, upper $(hkl)_8$ shown in Fig. 2.46 and lower $(hk\bar{1})_8$.

The tetragonal prisms of first and second order as well as the ditetragonal prism belonging to this class do not differ geometrically

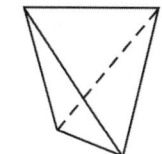

Fig. 2.40: Positive tetragonal disphenoid

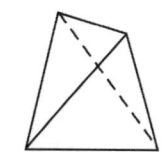

Fig. 2.41: Negative tetragonal disphenoid

Fig. 2.42: Positive tetragonal scalenohedron

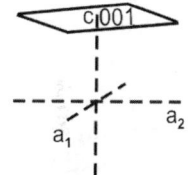

Fig. 2.43: Upper pedion

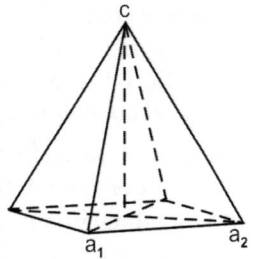

Fig. 2.44: Tetragonal pyramid of first order (upper)

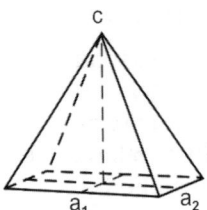

Fig. 2.45: Tetragonal pyramid of second order (upper)

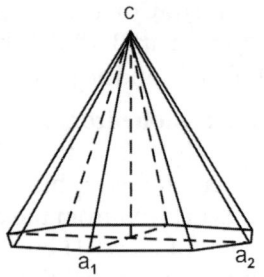

Fig. 2.46: Ditetragonal pyramid (upper)

from those of the $4/m\,2/m\,2/m$ class. However, they differ in molecular structure as exhibited by the development of striations, etc. The rare mineral diaboleite crystallises in this class.

2.3.4 Class 422 (Tetragonal Trapezohedral Class)

Symmetry elements: This class is characterised by the presence of an axis of 4-fold rotation, which is coincident with the vertical crystallographic axis and four axes of 2-fold rotation, two being coincident with horizontal crystallographic axes and other two bisect the angles between horizontal crystallographic axes. Mirror planes and centre of symmetry are absent.

Forms: The special form of this class is the tetragonal trapezohedron.

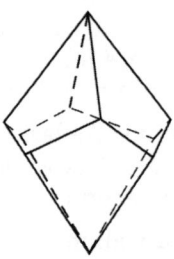

Fig. 2.47: Tetragonal trapezohedron (right)

Tetragonal trapezohedron: It is a solid bounded by eight trapezoid faces each of which intersects all the three crystallographic axes at unequal distances. There are two trapezohedra, right $(hkl)_8$ (Fig. 2.47) and left $(h\,\bar{k}\,l)_8$ (Fig. 2.48) corresponding to the ditetragonal dipyramid of the $4/m\,2/m\,2/m$ class. These two trapezohedra exhibit enantiomorphic relationship.

Other forms that may be present are basal pinacoid, first-, and second-order tetragonal prisms, ditetragonal prism and first- and second-order tetragonal dipyramids. Phosgenite is the only mineral crystallizing in this class.

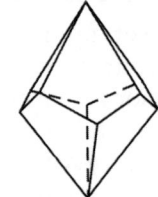

Fig. 2.48: Tetragonal trapezohedron (left)

2.3.5 Class 4/m (Tetragonal Dipyramidal Class)

Symmetry elements: This class is characterised by the presence of one axis of 4-fold rotation, which is coincident with the vertical crystallographic axis. In addition, there is one horizontal plane of symmetry containing the horizontal crystallographic axes. Centre of symmetry is also present.

Forms: Two special forms are present in this class. These are tetragonal prism of third order and tetragonal dipyramid of third order.

i. **Tetragonal prism of third order:** It is an open form having four rectangular faces each of which intersects both the horizontal crystallographic axes at unequal lengths. Corresponding to the ditetragonal prism of the $4/m\,2/m\,2/m$ class there are two prisms of third order, right $(hk0)_4$ shown in Fig. 2.49 and left $(kh0)_4$.

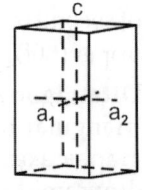

Fig. 2.49: Tetragonal prism of third order

ii. Tetragonal dipyramid of third order: It is a closed form bounded by eight scalene triangular faces each of which intersects all the three crystallographic axes at unequal lengths. Corresponding to the ditetragonal dipyramid of $4/m\,2/m\,2/m$ class there are two dipyramids of third order, right $(hkl)_8$ (Fig. 2.50) and left $(khl)_8$.

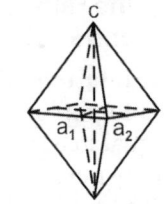

Fig. 2.50: Tetragonal dipyramid of third order

Other forms that may be present are basal pinacoid, first-, and second-order tetragonal prisms and first- and second-order tetragonal dipyramids. Representative minerals of this class are scheelite, powellite and members of the scapolite series.

2.3.6 Class $\overline{4}$ (Tetragonal Disphenoidal Class)

Symmetry elements: The vertical crystallographic axis is the axis of 4-fold rotoinversion. Plane and centre of symmetry are absent.

Forms: Tetragonal disphenoid of third order is the only special form present in this class.

i. Tetragonal disphenoid of third order: It is a solid bounded by four scalene triangular faces each of which intersects all the three crystallographic axes at unequal lengths. The form symbol is $(hkl)_4$. Corresponding to the ditetragonal dipyramid of the class $4/m\,2/m\,2/m$, there are four tetragonal disphenoids designated as right (positive and negative) and left (positive and negative). The right positive form is shown in Fig. 2.51.

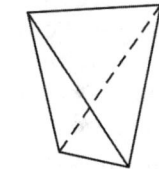

Fig. 2.51: Tetragonal disphenoid of third order (right-positive)

Other forms that may be present are basal pinacoid, first-, and second-order tetragonal prisms, ditetragonal prism and first- and second-order tetragonal dipyramids.

The rare mineral cahnite crystallizes in this class.

2.3.7 Class 4 (Tetragonal Pyramidal Class)

Symmetry elements: The vertical crystallographic axis is the only axis of 4-fold rotation. Plane and centre of symmetry are absent.

Forms: Tetragonal pyramid is the only special form present in this class.

i. **Tetragonal pyramid:** It is an open form having four scalene triangular faces each of which intersects all the three crystallographic axes at unequal lengths. The form symbol is $(hkl)_4$. Corresponding to the ditetragonal dipyramid of the class $4/m\,2/m\,2/m$, there are four tetragonal pyramids designated as positive (right- and left-handed) and negative (right- and left-handed). The positive right-handed form is shown in Fig. 2.52.

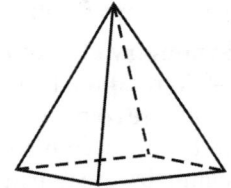

Fig. 2.52: Tetragonal pyramid (positive right handed)

Other forms that may be present are basal pinacoid, first- and second-order tetragonal prisms, ditetragonal prism and first- and second-order tetragonal dipyramids.

Wulfenite is the representative mineral crystallizing in this class.

2.4 IDENTIFICATION OF TETRAGONAL CRYSTAL FORMS

Four combination forms are shown in Figs 2.53–2.56. These are to be identified with respect to form, class and system.

2.4.1 The Form is a Combination of Two Sets of Faces, Triangular (A) and Rectangular (B) (Fig. 2.53)

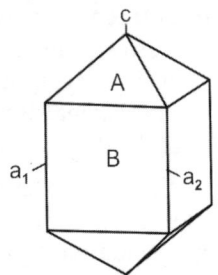

Axial relationship: The combination form can be referred to three crystallographic axes, out of which two horizontal (shown as a_1 and a_2) are equal in length and the vertical (c) is greater in length. All the three axes are perpendicular to each other. Thus, the form represented by Fig. 2.53 belongs to tetragonal system.

Fig. 2.53: Tetragonal combination form

Symmetry elements: The vertical crystallographic axis is the axis of 4-fold rotation. In addition, there are four axes of 2-fold rotation, two of which are coincident with the horizontal crystallographic axes a_1 and a_2 and other two bisect the angles between them. The planes of symmetry are five in number, three axial and two diagonal in position. Centre of symmetry is present. Thus, the form belongs to $4/m\,2/m\,2/m$ class.

Forms: There are two sets of faces in the model. The triangular faces (A) are eight in number, each intersects both the horizontal crystallographic axes at equal lengths and the vertical crystallographic axis at different length. Thus, the form symbol is $(hhl)_8$ and it is the tetragonal dipyramid of first order. The rectangular faces (B) are four in number, each intersects two horizontal crystallographic axes at equal lengths remaining parallel with the vertical crystallographic axis. The form symbol is $(110)_4$. It is the tetragonal prism of first order.

Conclusion: The model shown in Fig. 2.53 is a combination of dipyramid of first order and prism of first order of class $4/m\,2/m\,2/m$ (ditetragonal dipyramidal) of tetragonal system.

2.4.2 The Form is a Combination of Two Sets of Faces, Rectangular (A) and Hexagonal (B) (Fig. 2.54)

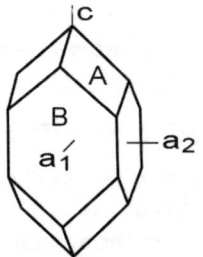

Axial relationship: The combination form can be referred to three crystallographic axes out of which two are horizontal, emerge at the midpoint of opposite hexagonal faces (shown as a_1 and a_2) and equal in length. The third one is vertical (c) and is greater in length. All the three axes are perpendicular to each other. Thus, the form represented by Fig. 2.54 belongs to tetragonal system.

Fig. 2.54: Tetragonal combination form

Symmetry elements: The vertical crystallographic axis is the axis of 4-fold rotation. In addition, there are four axes of 2-fold rotation, two of which are coincident with the horizontal crystallographic axes a_1 and a_2 and other two bisect the angles between horizontal crystallographic axes. The planes of symmetry are five in number, three axial and two diagonal in position. Centre of symmetry is present. Thus, the form belongs to $4/m\,2/m\,2/m$ class.

Forms: There are two sets of faces in the model. The rectangular faces (A) are eight in number, each intersects both the horizontal crystallographic axes at equal lengths and the vertical crystallographic axis at different length. The form symbol is $(hhl)_8$ and it is the tetragonal dipyramid of first order. The hexagonal faces (B) are four in number, each intersects one of the horizontal crystallographic axis remaining parallel with the other two crystallographic axes. The form symbol is $(100)_4$. It is the tetragonal prism of second order.

Conclusion: The model shown in Fig. 2.54 is a combination of dipyramid of first order and prism of second order of class $4/m\,2/m\,2/m$ (ditetragonal dipyramidal) of tetragonal system.

2.4.3 The Form is a Combination of four Sets of Faces, Hexagonal (A), Bigger Rectangular (B), Elongated Hexagonal (C) and Smaller Rectangular (D) (Fig. 2.55)

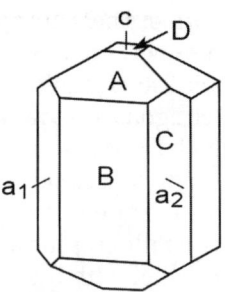

Axial relationship: The combination form can be referred to three crystallographic axes, two horizontal emerging at the midpoint of opposite elongated hexagonal faces (shown as a_1 and a_2) are equal in length and the vertical (c) is greater in length. All the three axes are perpendicular to each other. Thus, the form represented by Fig. 2.55 belongs to tetragonal system.

Fig. 2.55: Tetragonal combination form

Symmetry elements: The vertical crystallographic axis is the axis of 4-fold rotation. In addition, there are four axes of 2-fold rotation, two of which are coincident with the horizontal crystallographic axes a_1 and a_2 and other two bisect the angles between them. The planes of symmetry are five in number three axial and two diagonal in position. Centre of symmetry is present. Thus, the form belongs to $4/m\,2/m\,2/m$ class.

Forms: There are four sets of faces in the model. The hexagonal (A) faces are eight in number, each intersects both the horizontal crystallographic axes at equal lengths and the vertical crystallographic axis at different length. The form symbol is $(hhl)_8$ and it is the tetragonal dipyramid of first order. The bigger rectangular (B) faces are four in number, each intersects both the horizontal crystallographic axes at equal lengths remaining parallel with the vertical crystallographic axis. The form symbol is $(110)_4$. It is the tetragonal prism of first order. The elongated hexagonal (C) faces are four in number, each intersects one of the horizontal crystallographic axis remaining parallel with the other two crystallographic axes. The form symbol is $(100)_4$. It is the tetragonal prism of first order. The small rectangular (D) faces are two in number each intersects only the vertical crystallographic axis. The form symbol is $(001)_2$ and it is basal pinacoid.

Conclusion: The model shown in Fig. 2.55 is a combination of dipyramid of first order, prism of first order, prism of second order and basal pinacoid of ditetragonal dipyramidal class ($4/m\,2/m\,2/m$) of tetragonal system.

2.4.4 The Form is a Combination of Four Sets of Faces, Bigger Pentagonal (A), Octagonal (B), Smaller Pentagonal (C) and Rectangular (D) (Fig. 2.56)

Axial relationship: The combination form can be referred to three crystallographic axes, two horizontal emerging at the midpoint of opposite rectangular faces (shown as a_1 and a_2) are equal in length and the vertical (c) is greater in length. All the three axes are perpendicular to each other. Thus, the form represented by Fig. 2.56 belongs to tetragonal system.

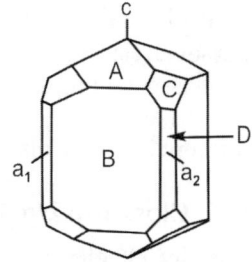

Symmetry elements: The vertical crystallographic axis is the axis of 4-fold rotation. In addition, there are four axes of 2-fold rotation, two of which are coincident with the horizontal crystallographic axes and other two bisect the angles between horizontal crystallographic axes. The planes of symmetry are five in number, three axial and two diagonal in position. Centre of symmetry is present. Thus, the form belongs to $4/m\,2/m\,2/m$ class.

Fig. 2.56: Tetragonal combination form

Forms: There are four sets of faces in the model. The bigger pentagonal (A) faces are eight in number, each intersects both the horizontal crystallographic axes at equal lengths and the vertical crystallographic axis at different length. Thus, the form symbol is $(hhl)_8$ and it is the tetragonal dipyramid of first order. The octagonal (B) faces are four in number, each intersects both the horizontal crystallographic axes at equal length remaining parallel with the vertical crystallographic axis. The form symbol is $(110)_4$. It is the tetragonal prism of first order. The smaller pentagonal (C) faces are eight in number, each intersects one of the horizontal crystallographic axis and the vertical crystallographic axis at different lengths remaining parallel with the other horizontal crystallographic axis. The form symbol is $(h0l)_8$. It is the tetragonal dipyramid of second order. The rectangular (D) faces are four in number each intersects one of the horizontal crystallographic axis remaining parallel with the other two crystallographic axes. The form symbol is $(100)_4$. It is the tetragonal prism of second order.

Conclusion: The model shown in Fig. 2.56 is a combination of dipyramid of first order, dipyramid of second order, prism of first order and prism of second order of ditetragonal dipyramidal class $(4/m\,2/m\,2/m)$ of tetragonal system.

2.5 HEXAGONAL SYSTEM

Axial relationship: Crystals belonging to the hexagonal system are referred to four crystallographic axes. Three of them lie in the horizontal plane, are equal in length with angles of 120° between their positive ends. These axes are mutually interchangeable and are designated as a_1, a_2 and a_3. The fourth axis, c, is vertical and is of different length. When properly oriented, the horizontal crystallographic

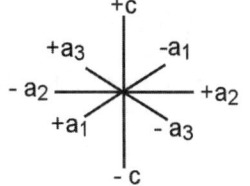

Fig. 2.57: Crystallographic axes of hexagonal system

axis, a_2, is left to right, positive at right hand side and negative at left hand side of the observer. The other two horizontal axes make 120° angles on either side of a_2. The positive end of a_1 is to the front and left while negative end of a_3 is to front and right. The orientation of the crystallographic axes and the method of their notation are shown in Fig. 2.57. The axial ratio (a:c) in case of beryl and quartz are 1:0.996 and 1:1.1 respectively. Corresponding to four crystallographic axes, four numbers (h, k, i and l) indicate the symbol of a face and the algebraic sum of the indices h, k, i is equal to zero.

Twelve classes are grouped under this system. These are $6/m\,2/m\,2/m$ (dihexagonal dipyramidal), $\bar{6}m2$ (ditrigonal dipyramidal), 6 mm (dihexagonal pyramidal), 622 (hexagonal trapezohedral), $6/m$ (hexagonal dipyramidal), $\bar{6}$ (trigonal dipyramidal), 6 (hexagonal pyramidal), $\bar{3}2/m$ (hexagonal scalenohedral), 3m (ditrigonal pyramidal), 32 (trigonal trapezohedral), $\bar{3}$ (rhombohedral) and 3 (trigonal pyramidal). According to some authors, the first seven classes belong to the hexagonal division of the hexagonal system while the remaining five classes belong to the rhombohedral or trigonal division of the hexagonal system (or trigonal system, according to some authors). Characteristic symmetry elements of each class and different forms available in them are discussed below.

2.5.1 Class 6/m 2/m 2/m (Dihexagonal Dipyramidal Class)

This class is also known as the *normal class of hexagonal system* as it possesses maximum symmetry among all the classes belonging to this system and *beryl type* after the characteristic mineral beryl that crystallizes in this class.

Symmetry elements: The vertical crystallographic axis is the axis of 6-fold rotation. There are six horizontal axes of 2-fold rotation, three of which are coincident with the horizontal crystallographic axes and other three bisect the angles between horizontal axes. There are seven mirror planes, one horizontal that contains all the three horizontal crystallographic axes and six vertical, three of which are axial each containing one of the horizontal crystallographic axis and the vertical crystallographic axis and other three bisect the angles between the vertical axial planes. Centre of symmetry is present.

Forms: Different forms belonging to this class are basal pinacoid, three types of prisms and corresponding three types of dipyramids.

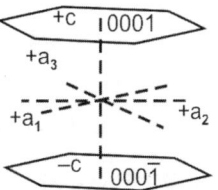

Fig. 2.58: Basal pinacoid

i. **Basal pinacoid (0001)$_2$:** The basal pinacoid (Fig. 2.58) is an open form having two horizontal faces, which remain parallel with the horizontal axes and intersect the vertical crystallographic axis at equal lengths. The form symbol is (0001)$_2$ and the two faces are 0001 and $000\bar{1}$. As it forms the base of the crystal, it is known as basal pinacoid.

ii. **Hexagonal prism of first order (10$\bar{1}$0)$_6$:** The hexagonal prism of first order (Fig. 2.59) is an open form that consists of six rectangular faces, each of which intersects two of the horizontal crystallographic axes at equal lengths and remains parallel with third horizontal and vertical crystallographic axes. The form symbol is (10$\bar{1}$0)$_6$. Six faces are 10$\bar{1}$0, 01$\bar{1}$0, $\bar{1}$100, $\bar{1}$010, 0$\bar{1}$10, and 1$\bar{1}$00.

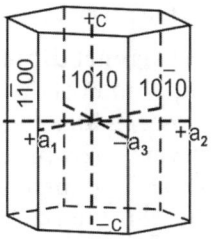

Fig. 2.59: Hexagonal prism of first order

iii. **Hexagonal prism of second order (11$\bar{2}$0)$_6$:** The hexagonal prism of second order (Fig. 2.60) is an open form that consists of six rectangular faces, each of which intersects two of the horizontal crystallographic axes at equal distances and the intermediate horizontal axis at half of this length remaining parallel with the vertical crystallographic axis. The form symbol is (11$\bar{2}$0)$_6$. Six faces are 11$\bar{2}$0, $\bar{1}$2$\bar{1}$0, $\bar{2}$110, $\bar{1}$ $\bar{1}$20, 1$\bar{2}$10, and 2$\bar{1}$ $\bar{1}$0.

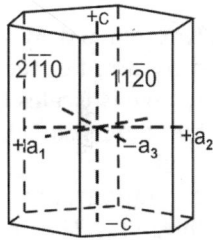

Fig. 2.60: Hexagonal prism of second order

iv. **Dihexagonal prism (hk$\bar{i}$0)$_{12}$:** The dihexagonal prism (Fig. 2.61) is an open form that consists of twelve rectangular faces, each of which intersects all the three horizontal crystallographic axes at different lengths remaining parallel with the vertical axis. The form symbol is (hk$\bar{i}$0)$_{12}$. This prism can be thought of as derived from the hexagonal prism of first order by development of two faces on each face of the hexagonal prism of first order. Depending upon the different intercepts made on the horizontal crystallographic axes a number of dihexagonal prisms exist but the most common one is (12$\bar{3}$0)$_{12}$.

The relation between the three prisms is shown in Fig. 2.62, which is a cross section in the horizontal plane of symmetry. Both the prisms of first and second order are geometrically alike and interchangeable with a rotation by 30° about the vertical

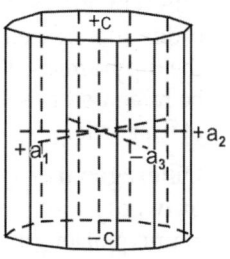

Fig. 2.61: Dihexagonal prism

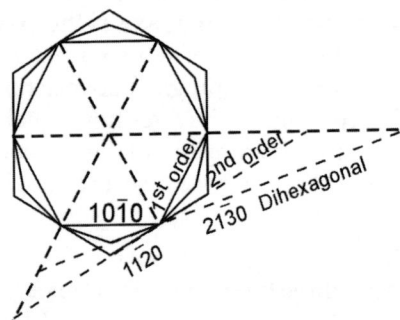

Fig. 2.62: Relationship between three types of prisms of hexagonal system

crystallographic axis. In case of prism of first order, the horizontal axes emerge at the midpoints of opposite vertical edges while in case of prism of second order, the horizontal axes are perpendicular to the prism faces at their midpoints.

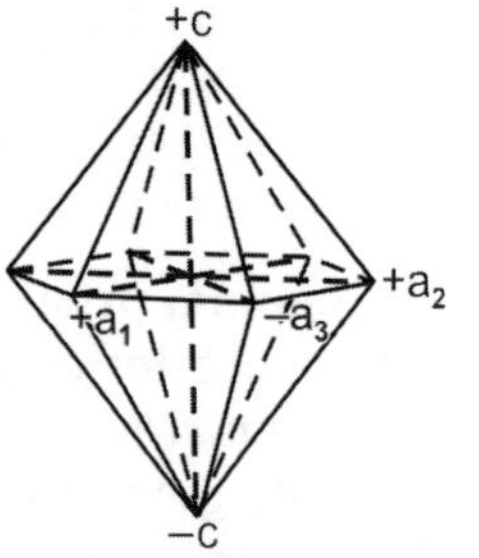

v. **Hexagonal dipyramid of first order $(h0\bar{h}l)_{12}$:** The hexagonal dipyramid of first order (Fig. 2.63) is a closed form having twelve isosceles triangular faces, each of which intersects two horizontal crystallographic axes at equal lengths and the vertical crystallographic axis at a different length remaining parallel with the third horizontal crystallographic axis. The form symbol is $(h0\bar{h}l)_{12}$. If a face of this dipyramid cuts the horizontal and the vertical crystallographic axes in the ratio equal to axial ratio, the form is called *unit* or *fundamental* form and the symbol becomes $(10\bar{1}1)_{12}$.

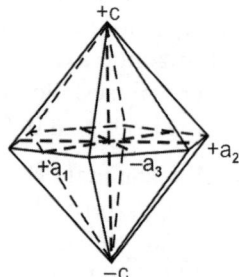

Fig. 2.63: Hexagonal dipyramid of first order

vi. **Hexagonal dipyramid of second order $(hh\overline{2h}l)_{12}$:** The hexagonal dipyramid of second order (Fig. 2.64) is a closed form having twelve isosceles triangular faces, each of which intersects two horizontal axes at equal distances and the third (the intermediate horizontal axis) at half of this distance and the vertical crystallographic axis at different length. The form symbol is $(hh\overline{2h}l)_{12}$. The unit dipyramid $(11\overline{2}1)_{12}$ is most common.

Similar to the tetragonal system, if only one dipyramid is present, it is set as first order. However, if two dipyramids of different orders are present, the dominant one is usually set as first order. In the orientation of a crystal, the prisms are subordinate to the dipyramids. Thus, an important prism with bigger faces may be relegated to second order by the presence of a small pyramid of first order.

Fig. 2.64: Hexagonal dipyramid of second order

vii. **Dihexagonal dipyramid $(hk\bar{i}l)_{24}$:** The dihexagonal dipyramid (Fig. 2.65) is a solid bounded by twenty-four scalene triangular faces each of which intersects all the crystallographic axes at unequal lengths. The form symbol is $(hk\bar{i}l)_{24}$. There are various dihexagonal dipyramids depending upon the different intersections on the crystallographic axes but the form $(21\bar{3}1)_{24}$ is most common.

Important minerals crystallizing in this class are beryl, molybdenite, pyrrhotite, nicolite, covellite, tridymite, etc.

2.5.2 Class $\bar{6}$ m2 (Ditrigonal Dipyramidal Class)

Symmetry elements: The vertical crystallographic axis is the axis of 6-fold rotoinversion. Since $\bar{6}$ is equivalent to 3+m, the c-axis appears to be an axis of 3-fold rotation. There are three axes of 2-fold rotation, which bisect the angles between three horizontal crystallographic axes. In addition, there are four mirror planes, one being horizontal and others are diagonal each containing the c-axis and one of the horizontal symmetry axis. Centre of symmetry is absent.

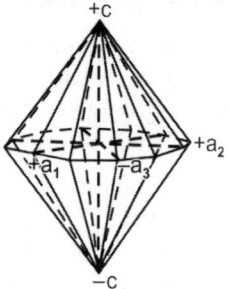

Fig. 2.65: Dihexagonal dipyramid

Forms: Trigonal prism, ditrigonal prism, trigonal dipyramid and ditrigonal dipyramid are the special forms present in this class.

i. **Trigonal prism:** It is an open form consisting of three vertical rectangular faces each of which intersects two of the horizontal crystallographic axes at equal lengths remaining parallel with the third horizontal and the vertical crystallographic axes. Corresponding to the hexagonal prism of first order there are two trigonal prisms designated as positive $(10\bar{1}0)_3$ and negative $(01\bar{1}0)_3$. The positive form is shown in Fig. 2.66.

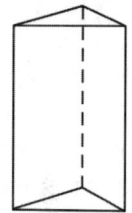

Fig. 2.66: Trigonal prism

ii. **Ditrigonal prism:** It is an open form having six vertical rectangular faces each of which intersects all the three horizontal crystallographic axes at different lengths remaining parallel with the vertical crystallographic axis. Thus, there are two ditrigonal prisms corresponding to the dihexagonal prism of class $6/m\,2/m\,2/m$ designated as positive $(hk\bar{i}0)_6$ (Fig. 2.67) and negative $(kh\bar{i}0)_6$.

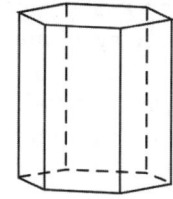

Fig. 2.67: Ditrigonal prism

iii. **Trigonal dipyramid:** It is a double three-faced pyramid having six isosceles triangular faces each of which intersects two of the horizontal crystallographic axes at equal lengths and the vertical axis at different length remaining parallel with the third horizontal crystallographic axis. Corresponding to the hexagonal dipyramid of first order there are two trigonal dipyramids, positive $(10\bar{1}1)_6$ and negative $(01\bar{1}1)_6$. The positive form is shown in Fig. 2.68.

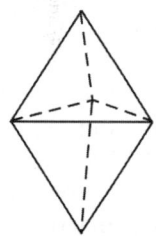

Fig. 2.68: Trigonal dipyramid

iv. **Ditrigonal dipyramid:** It is a closed form bounded by twelve scalene triangular faces (Fig. 2.69). Each face intersects all the four crystallographic axes at unequal lengths. It is thought to have been derived from the dihexagonal dipyramid of class $6/m\,2/m\,2/m$ by development of alternate pairs of faces. Thus, there are two complementary forms designated as positive $(hk\bar{i}l)_{12}$ and negative $(kh\bar{i}l)_{12}$.

Other forms that may be present are basal pinacoid, hexagonal prism of second-order and hexagonal dipyramid of second-order. Benitoite is the only mineral representative of this class.

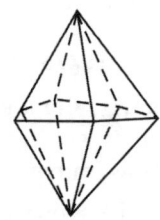

Fig. 2.69: Ditrigonal dipyramid

2.5.3 Class 6 mm (Dihexagonal Pyramidal Class)

Symmetry elements: This class is characterised by the presence of an axis of 6-fold rotation coincident with the vertical crystallographic axis and six vertical mirror planes, three containing the horizontal crystallographic axes and remaining three are in diagonal position. Centre of symmetry is absent.

Forms: The forms of the dihexagonal pyramidal class are similar to those of the dihexagonal dipyramidal class, but, as the horizontal mirror plane is absent, different forms appear at the top and bottom of the crystal. The basal pinacoid is dissected into two pedions, upper $(0001)_1$ and lower $(000\overline{1})_1$. The dipyramids are segmented into pyramids occurring at the extremities of the vertical crystallographic axis. The hexagonal dipyramid of first order occurs in two forms, upper $(h0\overline{h}l)_6$ and lower $(h0\overline{h}\overline{1})_6$; hexagonal dipyramid of second order occurs in two forms, upper $(hh\overline{2h}l)_6$ and lower $(hh\overline{2h1})_6$ and the dihexagonal dipyramid occurs as dihexagonal pyramid upper $(hk\overline{i}l)_{12}$ shown in Fig. 2.70 and lower $(hk\overline{i}\,\overline{1})_{12}$ forms. In addition to these forms, first- and second-order hexagonal prisms and the dihexagonal prism are also present.

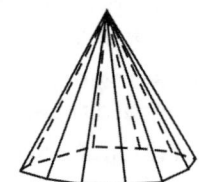

Fig. 2.70: Dihexagonal pyramid (upper)

Wurtzite, greenockite and zincite are the common minerals crystallizing in this class.

2.5.4 Class 622 (Hexagonal Trapezohedral Class)

Symmetry elements: The vertical crystallographic axis is the axis of 6-fold rotation and there are six axes of 2-fold rotation out of which three are coincident with the horizontal crystallographic axes and remaining three are in diagonal position. Mirror plane and centre of symmetry are absent.

Forms: Hexagonal trapezohedron is the only special form present in this class. It is a solid bounded by twelve trapezium-shaped faces, six above and six below. Each face intersects all the four crystallographic axes at unequal lengths. Corresponding to the dihexagonal dipyramid of the class $6/m\ 2/m\ 2/m$, there are two trapezohedrons designated as right $(hk\overline{i}l)_{12}$ and left $(ih\overline{k}l)_{12}$. These two enantiomorphic forms are shown in Figs 2.71 and 2.72 respectively. Other forms that may be present are pinacoid, first- and second-order hexagonal prisms, dipyramids and dihexagonal prism.

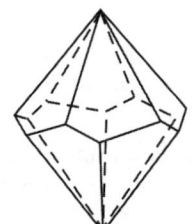

Fig. 2.71: Hexagonal trapezohedron (right)

High-temperature quartz (β-quartz) and kalsilite are the important minerals crystallizing in this class.

2.5.5 Class 6/m (Hexagonal Dipyramidal Class)

Symmetry elements: An axis of 6-fold rotation coincident with the vertical crystallographic axis, one horizontal axial mirror plane and centre of symmetry are the symmetry elements of this class.

Forms: Hexagonal prism of third order and hexagonal dipyramid of third order are the special forms of this class.

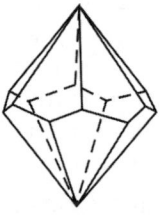

Fig. 2.72: Hexagonal trapezohedron (left)

i. **Hexagonal prism of third order:** It is an open form having six rectangular faces each of which intersects all the three horizontal crystallographic axes at unequal lengths remaining parallel with the vertical crystallographic axis (Fig. 2.73). Corresponding to the dihexagonal prism of the class $6/m\,2/m\,2/m$, there are two hexagonal prisms of third order designated as positive $(hk\bar{i}0)_6$ and negative $(kh\bar{i}0)_6$.

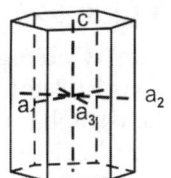

Fig. 2.73: Hexagonal prism of third order

ii. **Hexagonal dipyramid of third order:** It is a closed form bounded by twelve scalene triangular faces each of which intersects all the four crystallographic axes at unequal lengths. Corresponding to the dihexagonal dipyramid of the class $6/m\,2/m\,2/m$, there are two hexagonal dipyramids of third order designated as positive $(hk\bar{i}l)_{12}$ and negative $(kh\bar{i}l)_{12}$. The positive hexagonal dipyramid is shown in Fig. 2.74.

Other forms that may be present in this class are basal pinacoid, first- and second-order hexagonal prisms, dipyramids and dihexagonal prism. Apatite is the important mineral crystallizing in this class.

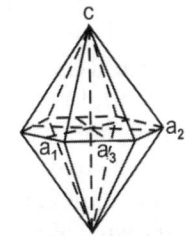

Fig. 2.74: Hexagonal dipyramid of third order

2.5.6 Class $\bar{6}$ (Trigonal Dipyramidal Class)

Symmetry elements: The vertical crystallographic axis is an axis of 6-fold rotoinversion, which is equivalent to a 3-fold axis of rotation with a symmetry plane at right angle to it. Centre of symmetry is absent.

Forms: The symmetry does not permit hexagonal forms. Thus, instead of first- and second-order hexagonal prisms, there are two first-order and two second-order trigonal prisms. Similarly, in place of first- and second-order hexagonal dipyramids there are two trigonal dipyramids. Corresponding to the 24 faces of the dihexagonal dipyramid, there are four trigonal dipyramids each having six faces, three above and three below. One such dipyramid is shown in Fig. 2.75. The basal pinacoid may be present. This class has no mineral representative.

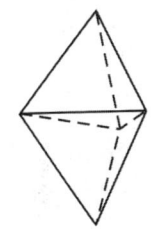

Fig. 2.75: Trigonal dipyramid

2.5.7 Class 6 (Hexagonal Pyramidal Class)

Symmetry elements: The class is characterised by the presence of one axis of 6-fold rotation coincident with the vertical crystallographic axis. Mirror plane and centre of symmetry are absent.

Forms: Since the vertical crystallographic axis is an axis of 6-fold rotation and horizontal plane of symmetry is absent, the pyramids are differentiated into forms having six faces at the upper and lower extremities of the vertical crystallographic axis. Thus, the first- and second-order hexagonal dipyramids are divided into two forms each. Similarly, the dihexagonal dipyramid is segmented into four forms, upper positive and negative and lower positive and negative. One such hexagonal pyramid is shown in Fig. 2.76. The dihexagonal prism is segmented into two hexagonal prisms, $(hk\bar{i}0)_6$ and $(kh\bar{i}0)_6$. The basal pinacoid and first- and second-order hexagonal prisms may be present.

Nepheline is the chief mineral representative of this class.

Fig. 2.76: Hexagonal pyramid

2.5.8 Class $\bar{3}$ 2/m (Hexagonal Scalenohedral Class)

Symmetry elements: The vertical crystallographic axis is the axis of 3-fold rotoinversion and the three horizontal crystallographic axes are the axes of 2-fold rotation. There are three vertical mirror planes bisecting the angles between the horizontal axes. Centre of symmetry is present.

Forms: Two special forms are available in $\bar{3}$ 2/m class. These are rhombohedron and scalenohedron.

i. **Rhombohedron:** The rhombohedron is a solid bounded by six rhomb-shaped faces each of which intersects two of the horizontal crystallographic axes at equal lengths and the vertical crystallographic axis at a different length remaining parallel with the third horizontal crystallographic axis. The form symbol is $(h0\bar{h}l)_6$. There are two rhombohedrons corresponding to the first order hexagonal dipyramid of the 6/m 2/m 2/m class, positive $(h0\bar{h}l)_6$ and negative $(0h\bar{h}l)_6$ shown in Fig. 2.77 and Fig. 2.78 respectively. When properly oriented, the positive rhombohedron has one of its face and the negative rhombohedron has one of its edge towards the observer. The symbols of the unit positive rhombohedron is $(10\bar{1}1)_6$ and unit negative rhombohedron is $(01\bar{1}1)_6$.

ii. **Hexagonal scalenohedron:** The hexagonal scalenohedron is a closed form consisting of twelve scalene triangular faces. Each face intersects all the three horizontal crystallographic axes at unequal lengths and the vertical crystallographic axis at a different length. Corresponding to the dihexagonal dipyramid of the 6/m 2/m 2/m class, there are two scalenohedrons designated as positive $(hk\bar{i}l)_{12}$ and negative $(kh\bar{i}l)_{12}$. The scalenohedron looks like a dipyramid but is distinguished from the dipyramid by the zigzag appearance of the middle edges. It is in the positive position when the angle between the upper and lower faces point down towards the observer (Fig. 2.79) and in the negative position when the angle points up (Fig. 2.80).

Other available forms are basal pinacoid, first- and second-order hexagonal prisms, dihexagonal prism and hexagonal dipyramid of second order. Representative minerals of this class are calcite, corundum, hematite, brucite, etc.

2.5.9 Class 3m (Ditrigonal Pyramidal Class)

Symmetry elements: The vertical crystallographic axis is the axis of 3-fold rotation and there are three mirror planes each of which contains the vertical axis and bisects the angle between the horizontal axes. Centre of symmetry is absent.

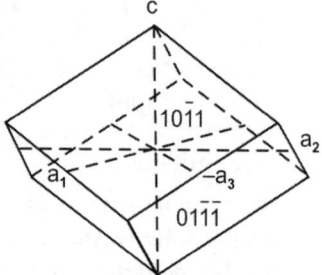

Fig. 2.77: Rhombohedron (positive)

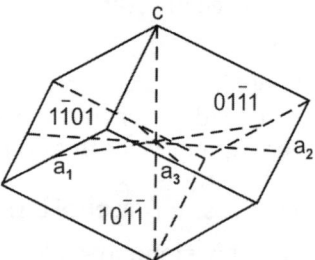

Fig. 2.78: Rhombohedron (negative)

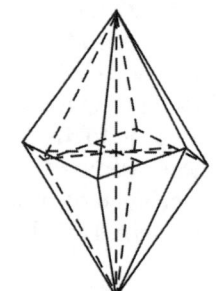

Fig. 2.79: Hexagonal scalenohedron (positive)

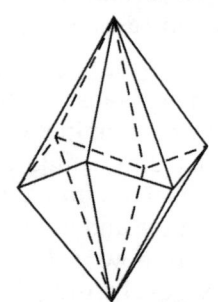

Fig. 2.80: Hexagonal scalenohedron (negative)

Forms: The forms are similar to those of the hexagonal scalenohedral class but with only half of the number of faces. Due to the lack of 2-fold rotational axis and horizontal plane of symmetry, the faces at the top of the crystals belong to different forms from those at the bottom. Basal pinacoid is differentiated into two pedions. Prism of first-order is represented by two trigonal prisms $(10\bar{1}0)_3$ and $(01\bar{1}0)_3$; dihexagonal prism is differentiated into two ditrigonal prisms $(hk\bar{i}0)_6$ and $(kh\bar{i}0)_6$. Positive and negative rhombohedrons are represented by four trigonal pyramids, two at the top $(h0\bar{h}l)_3$ and $(0h\bar{h}l)_3$ and two at the bottom $(0h\bar{h}\bar{l})_3$ and $(h0\bar{h}\bar{l})_3$. Similarly, positive and negative scalenohedrons are represented by four ditriginal pyramids with symbols $(hk\bar{i}l)_6$, $(kh\bar{i}l)_6$, $(hk\bar{i}\bar{l})_6$ and $(kh\bar{i}\bar{l})_6$. The ditriginal pyramid $(hk\bar{i}l)_6$ is shown in Fig. 2.81. Other forms that may be present are second-order prism and dipyramid.

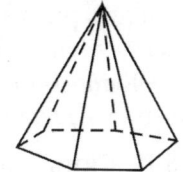

Fig. 2.81: Ditrigonal pyramid

Tourmaline is the most common mineral crystallizing in this class in addition to pyrargyrite, proustite and alunite.

2.5.10 Class 32 (Trigonal Trapezohedral Class)

Symmetry elements: The vertical crystallographic axis is the axis of 3-fold rotation and the three horizontal crystallographic axes are axes of 2-fold rotation. Plane and centre of symmetry are absent.

Forms: The characteristic form of this class is the trigonal trapezoheron. It is a closed form bounded by six trapezoid faces, each intersecting all the crystallographic axes at unequal lengths. Corresponding to the dihexagonal dipyramid, there are four trapezohedrons designated as positive right $(hk\bar{i}l)_6$ (Fig. 2.82), positive left $(i\bar{k}\bar{h}l)_6$ (Fig. 2.83), negative right $(\bar{k}ih\bar{l})_6$ and negative left $(kh\bar{i}l)_6$. The right and left forms are enantiomorphic pairs. Other forms that may be present are basal pinacoid, first-order hexagonal prism, ditrigonal prisms and rhombohedrons. The second-order hexagonal prism is differentiated into two trigonal prisms $(11\bar{2}0)_3$ and $(2\bar{1}\bar{1}0)_3$. Likewise the second-order hexagonal dipyramids are represented by two trigonal dipyramids $(hh\overline{2h}l)_6$ and $(2h\bar{h}\bar{h}l)_6$.

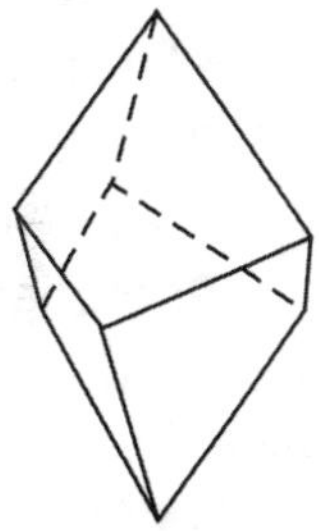

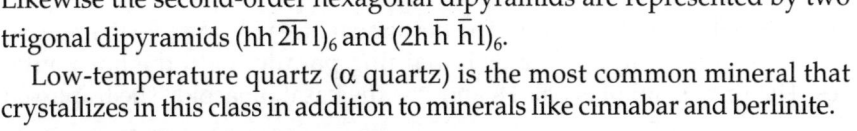

Fig. 2.82: Trigonal trapezohedron (positive right)

Low-temperature quartz (α quartz) is the most common mineral that crystallizes in this class in addition to minerals like cinnabar and berlinite.

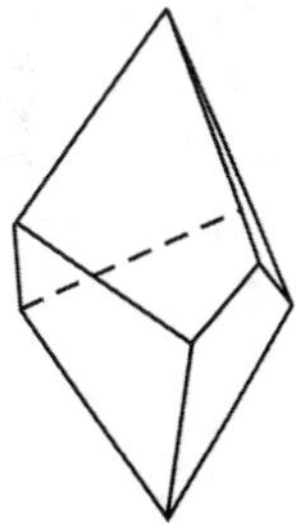

2.5.11 Class $\bar{3}$ (Rhombohedral Class)

Symmetry elements: The vertical axis is the axis of 3-fold rotoinversion, which is equivalent to a 3-fold rotation axis and centre of symmetry. Plane of symmetry is absent.

Fig. 2.83: Trigonal trapezohedron (positive left)

Forms: This class is characterised by three distinct types of rhombohedrons. The rhombohedrons of the class $\bar{3}\,2/m$, which are derived from the hexagonal dipyramid of first-order, are known as first-order rhombohedrons. The second order rhombohedrons are derived from the hexagonal dipyramid of second order by development of half of the number of faces. The third-order rhombohedron has the general symbol of $(hk\bar{i}l)_6$, and may be derived from the dihexagonal dipyramid by taking one quarter of the faces. One of this rhombohedron is shown in Fig. 2.84. The dihexagonal prism is differentiated into two hexagonal prisms $(hk\bar{1}0)_6$ and $(i\bar{k}\bar{h}0)_6$. Other forms belonging to this class are basal pinacoid and first- and second-order hexagonal prisms.

Dolomite is the most common mineral crystallizing in this class in addition to minerals like ilmenite, willemite and phenacite.

2.5.12 Class 3 (Trigonal Pyramidal Class)

Symmetry elements: The vertical crystallographic axis is the only axis of 3-fold rotation. Plane and centre of symmetry are absent.

Forms: The general form of this class is the trigonal pyramid (Fig. 2.85), which is an open form having three faces. There are eight trigonal pyramids corresponding to the 24 faces of the dihexagonal dipyramid. In addition, there are four first-order and four second-order trigonal pyramids. Pedions and several different trigonal prisms may be present.

Possibly the mineral gratonite crystallizes in this class.

2.6 IDENTIFICATION OF HEXAGONAL CRYSTAL FORMS

Six combination forms are shown in Figs 2.86–2.91. These are to be identified with respect to form, class and system.

2.6.1 The Form is a Combination of Three Sets of Faces, Hexagonal (A), Trapezoid (B) and Rectangular (C) (Fig. 2.86)

Axial relationship: The combination form can be referred to four crystallographic axes, out of which three are horizontal, equal in length, a_2 extends from right to left, a_1 and a_3 make 120° angles with a_2 as shown in Fig. 2.86. The vertical, c-axis, is different in length. Thus, the form represented by Fig. 2.86 belongs to hexagonal system.

Symmetry elements: The vertical crystallographic axis is the axis of 6-fold rotation; there are six axes of 2-fold rotation, three are coincident with the horizontal crystallographic axes and other three bisect the angles between horizontal crystallographic axes. There are seven mirror planes, one horizontal and six vertical. Centre of symmetry is present. Thus, the form belongs to dihexagonal dipyramidal class $(6/m\,2/m\,2/m)$.

Forms: There are three sets of faces in the model. The hexagonal faces (A) are two in number, each intersects the vertical crystallographic axis remaining parallel with the horizontal crystallographic axes. The form symbol is $(0001)_2$ and it is the basal pinacoid. The trapezoid

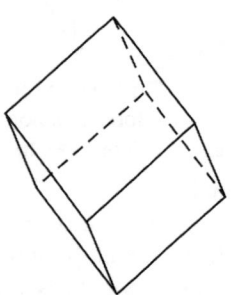

Fig. 2.84: Rhombohedron (positive right)

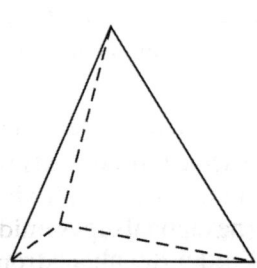

Fig. 2.85: Trigonal pyramid

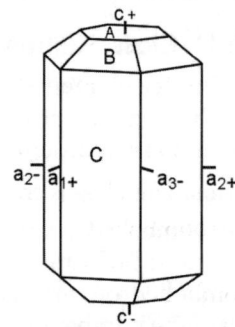

Fig. 2.86: Hexagonal combination form

faces (B) are 12 in number, each of which intersects two of the horizontal crystallographic axes at equal lengths and the vertical crystallographic axis at a different length remaining parallel with the third horizontal crystallographic axis. The form symbol is $(h0\bar{h}l)_{12}$ and it is the hexagonal dipyramid of first order. The rectangular faces (C) are six in number, each intersects two of the horizontal crystallographic axes at equal lengths remaining parallel with the intermediate horizontal and vertical crystallographic axes. The form symbol is $(10\bar{1}0)_6$ and it is the hexagonal prism of first order.

Conclusion: The model shown in Fig. 2.86 is a combination of basal pinacoid, prism of first order and dipyramid of first order of the dihexagonal dipyramidal class (6/m 2/m 2/m) of the hexagonal system.

2.6.2 The Form is a Combination of Three Sets of Faces, Hexagonal (A), Triangular (B) and Nine-sided (C) (Fig. 2.87)

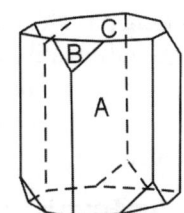

Axial relationship: The combination form can be referred to four crystallographic axes, out of which three are horizontal, equal in length and one vertical, which is different in length. Thus, the form represented by Fig. 2.87 belongs to hexagonal system.

Symmetry elements: The vertical crystallographic axis is the axis of 3-fold rotoinversion and the three horizontal crystallographic axes are axes of 2-fold rotation. There are three vertical mirror planes bisecting the angles between the horizontal axes. Centre of symmetry is present. Thus, the form belongs to hexagonal scalenohedral class ($\bar{3}$ 2/m).

Fig. 2.87: Trigonal combination form

Forms: There are three sets of faces in the model. The hexagonal faces (A) are six in number, each intersects two of the horizontal crystallographic axes at equal lengths and the intermediate horizontal axis at half of this length remaining parallel with the vertical crystallographic axis. The form symbol is $(11\bar{2}0)_6$. It is hexagonal prism of second order. The triangular faces (B) are six in number, each of which intersects two of the horizontal crystallographic axes at equal lengths and the vertical crystallographic axis at a different length remaining parallel with the third horizontal crystallographic axis. The form symbol is $(h0\bar{h}l)_6$ and it is the positive rhombohedron. The nine-sided faces are two in number, each intersects the vertical crystallographic axis remaining parallel with the horizontal crystallographic axes. The form symbol is $(0001)_2$ and it is the basal pinacoid.

Conclusion: The model shown in Fig. 2.87 is a combination of hexagonal prism of second order, positive rhombohedron and the basal pinacoid of the hexagonal scalenohedral class ($\bar{3}$ 2/m) of the hexagonal system.

2.6.3 The Form is a Combination of Three Sets of Faces, Hexagonal (A), Octagonal (B) and Triangular (C) (Fig. 2.88)

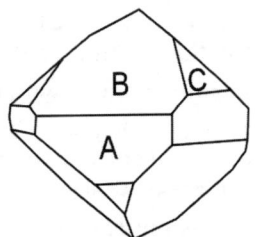

Axial relationship: The combination form can be referred to four crystallographic axes, out of which three are horizontal, equal in length and one vertical, which is different in length. Thus, the form represented by Fig. 2.88 belongs to hexagonal system.

Symmetry elements: The vertical crystallographic axis is the axis of 3-fold rotoinversion and the three horizontal crystallographic axes are the axes of 2-fold rotation. There are three vertical mirror planes

Fig. 2.88: Trigonal combination form

bisecting the angles between the horizontal axes. Centre of symmetry is present. Thus, the form belongs to hexagonal scalenohedral class ($\bar{3}\,2/m$).

Forms: There are three sets of faces in the model. The hexagonal faces (A) are six in number, each intersects two of the horizontal crystallographic axes at equal lengths remaining parallel with the intermediate horizontal axis and the vertical crystallographic axis. The form symbol is $(10\bar{1}0)_6$. It is hexagonal prism of first order. The octagonal faces (B) are six in number, each of which intersects two of the horizontal crystallographic axes at equal lengths and the vertical crystallographic axis at a different length remaining parallel with the third horizontal crystallographic axis. The form symbol is $(h0\bar{h}l)_6$ and it is the positive rhombohedron. The triangular faces (C) are six in number, each of which intersects two of the horizontal crystallographic axes at equal lengths and the vertical crystallographic axis at a different length remaining parallel with the third horizontal crystallographic axis. The form symbol is $(0h\bar{h}l)_6$ and it is the negative rhombohedron.

Conclusion: The model shown in Fig. 2.88 is a combination of hexagonal prism of first order and positive and negative rhombohedrons of hexagonal scalenohedral class ($\bar{3}\,2/m$) of the hexagonal system.

2.6.4 The Form is a Combination of Rectangular (A), Two Types of Trapezoids (B and C) and Hexagonal (D) Faces (Fig. 2.89)

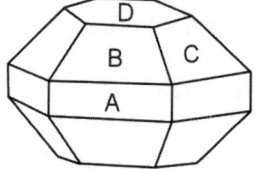

Fig. 2.89: Trigonal combination form

Axial relationship: The combination form can be referred to four crystallographic axes, out of which three are horizontal, equal in length and one vertical, which is different in length. Thus, the form represented by Fig. 2.89 belongs to hexagonal system.

Symmetry elements: The vertical crystallographic axis is the axis of 3-fold rotoinversion and the three horizontal crystallographic axes are the axes of 2-fold rotation. There are three vertical mirror planes bisecting the angles between the horizontal axes. Centre of symmetry is present. Thus, the form belongs to hexagonal scalenohedral class ($\bar{3}\,2/m$).

Forms: There are four sets of faces in the model. The rectangular faces (A) are six in number, each intersects two of the horizontal crystallographic axes at equal lengths remaining parallel with the intermediate horizontal axis and the vertical crystallographic axis. The form symbol is $(10\bar{1}0)_6$. It is hexagonal prism of first order. The trapezoid faces (B) are six in number, each of which intersects two of the horizontal crystallographic axes at equal lengths and the vertical crystallographic axis at different length remaining parallel with the third horizontal crystallographic axis. The form symbol is $(h0\bar{h}l)_6$ and it is the positive rhombohedron. The other trapezoid faces (C) are six in number, each of which intersects two of the horizontal crystallographic axes at equal lengths and the vertical crystallographic axis at a different length remaining parallel with the third horizontal crystallographic axis. The form symbol is $(0h\bar{h}l)_6$ and it is the negative rhombohedron. The hexagonal faces are two in number, each intersects the vertical crystallographic axis remaining parallel with the horizontal crystallographic axes. The form symbol is $(0001)_2$ and it is the basal pinacoid.

Conclusion: The model shown in Fig. 2.89 is a combination of hexagonal prism of first order, basal pinacoid and positive and negative rhombohedrons of hexagonal scalenohedral class ($\bar{3}\,2/m$) of the hexagonal system.

2.6.5 The Form is a Combination of Two Sets of Faces, Bigger Rectangular (A) and Smaller Rectangular (B) (Figure 2.90)

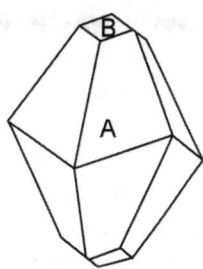

Fig. 2.90: Trigonal combination form

Axial relationship: The combination form can be referred to four crystallographic axes, out of which three are horizontal, equal in length and one vertical, which is different in length. Thus, the form represented by Fig. 2.90 belongs to hexagonal system.

Symmetry elements: The vertical crystallographic axis is the axis of 3-fold rotoinversion and the three horizontal crystallographic axes are the axes of 2-fold rotation. There are three vertical mirror planes bisecting the angles between the horizontal axes. Centre of symmetry is present. Thus, the form belongs to hexagonal scalenohedral class ($\bar{3}\,2/m$).

Forms: There are two sets of faces in the model. The bigger rectangular faces (A) are twelve in number, each intersects all the three horizontal crystallographic axes at unequal lengths and the vertical axis at different length. The form symbol is $(hk\bar{i}l)_{12}$ and it is the positive hexagonal scalenohedron. The smaller rectangular faces are six in number, each of which intersects two of the horizontal crystallographic axes at equal lengths and the vertical crystallographic axis at a different length remaining parallel with the third horizontal crystallographic axis. The form symbol is $(h0\bar{h}l)_6$ and it is the positive rhombohedron.

Conclusion: The model shown in Fig. 2.90 is a combination of positive hexagonal scalenohedron and positive rhombohedron of $\bar{3}\,2/m$ class (hexagonal scalenohedral) of the hexagonal system.

2.6.6 The Form is a Combination of Three Sets of Faces, Hexagonal (A), Pentagonal (B) and Rectangular (C) (Fig. 2.91)

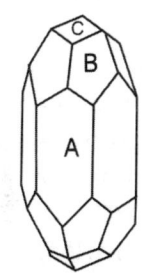

Fig. 2.91: Trigonal combination form

Axial relationship: The combination form can be referred to four crystallographic axes, out of which three are horizontal, equal in length and one vertical, which is different in length. Thus, the form represented by Fig. 2.91 belongs to hexagonal system.

Symmetry elements: The vertical crystallographic axis is the axis of 3-fold rotoinversion and the three horizontal crystallographic axes are axes of 2-fold rotation. There are three vertical mirror planes bisecting the angles between the horizontal axes. Centre of symmetry is present. Thus, the form belongs to hexagonal scalenohedral class ($\bar{3}\,2/m$).

Forms: There are three sets of faces in the model. The hexagonal faces (A) are six in number, each intersects two of the horizontal crystallographic axes at equal lengths remaining parallel with the intermediate horizontal and the vertical crystallographic axes. The form symbol is $(10\bar{1}0)_6$. It is hexagonal prism of first order. The pentagonal faces (B) are twelve in number, each intersects all the three horizontal crystallographic axes at unequal lengths and the vertical axis at different length. The form symbol is $(hk\bar{i}l)_{12}$ and it is the positive hexagonal scalenohedron. The smaller rectangular faces (C) are six in number, each of which intersects two of the horizontal crystallographic axes at equal lengths and the vertical crystallographic axis at a different length remaining parallel with the third horizontal crystallographic axis. The form symbol is $(h0\bar{h}l)_6$ and it is the positive rhombohedron.

Conclusion: The model shown in Fig. 2.91 is a combination of hexagonal prism of first order, positive hexagonal scalenohedron and positive rhombohedron of $\bar{3}\,2/m$ class (hexagonal scalenohedral) of the hexagonal system.

2.7 ORTHORHOMBIC SYSTEM

Axial relationship: Crystals belonging to the orthorhombic system are referred to three crystallographic axes, which are unequal in length and perpendicular to each other. They are designated as a, b and c. In orienting orthorhombic crystals, the convention is to make c < a < b. In proper orientation, the c-axis is vertical, positive at top and negative at bottom; b-axis runs from left to right, positive at the right hand side and negative at the left hand side and the a-axis is front to back, positive at the observer side and negative at the back of the crystal. Proper orientation of the crystallographic axes and the method of their notation are shown in Fig. 2.92. Previously the a-axis was called the brachy (short) axis and the b-axis the macro (long) axis. But this relationship is not universally true. The axial ratio (a:b:c) for sillimanite and barite are 0.98: 1: 0.75 and 0.815: 1: 1.31 respectively.

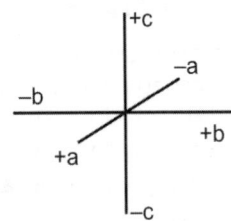

Fig. 2.92: Crystallographic axes of orthorhombic system

Three classes are grouped under this system. These are $2/m\,2/3\,2/m$ (rhombic dipyramidal), mm2 (rhombic pyramidal) and 222 (rhombic disphenoidal). Characteristic symmetry elements of each class and different forms available in them are described below.

2.7.1 Class 2/m 2/m 2/m (Rhombic Dipyramidal Class)

This class is also known as the *normal class of orthorhombic system* as it shows maximum symmetry among all the classes belonging to this system and *barite type* after the characteristic mineral barite that crystallizes in this class.

Symmetry elements: Three crystallographic axes are the axes of 2-fold rotation. There are three axial mirror planes each containing two of the crystallographic axes. Centre of symmetry is present.

Forms: Pinacoids, prisms and dipyramid are the characteristic forms of this class.

   i. **Front or a-pinacoid (100)₂:** It is an open form having two faces each of which intersects a-crystallographic axis remaining parallel with other two crystallographic axes (Fig. 2.93). The form symbol is $(100)_2$ and two faces are 100 and $\bar{1}\,00$. Its old name is macropinacoid, as both the faces are parallel with macro (b) axis. It is also known as front-pinacoid as one of the face remains in front of the crystal.

   ii. **Side or b-pinacoid (010)₂:** It is an open form having two faces each of which intersects b-crystallographic axis remaining parallel with other two crystallographic axes (Fig. 2.93). The form symbol is $(010)_2$ and two faces are 010 and $0\,\bar{1}\,0$. Its old name is brachypinacoid, as both the faces are parallel with brachy (a) axis. It is also known as side-pinacoid as both the faces remain to the right and left sides of the crystal.

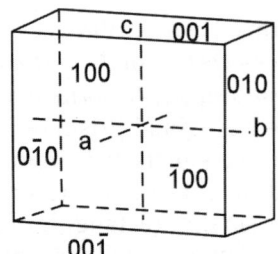

Fig. 2.93: a-, b- and c-pinacoids of orthorhombic system

iii. Basal or c-pinacoid (001)$_2$: It is an open form having two faces each of which intersects c-crystallographic axis remaining parallel with other two crystallographic axes (Fig. 2.93). The form symbol is (001)$_2$ and two faces are 001 and 00$\overline{1}$. It is also known as basal pinacoid as one of the face forms the base of the crystal.

Figure 2.93 shows a combination of a-, b- and c-pinacoids.

iv. First-order prism (0kl)$_4$: It is an open form having four faces each of which remains parallel with a-crystallographic axis and intersects b- and c-crystallographic axes at unequal lengths. The form symbol is (0kl)$_4$ and the unit form symbol is (011)$_4$. The first order prism is shown in Fig. 2.94. Its old name is brachydome, as the four faces are parallel with brachy (a) axis.

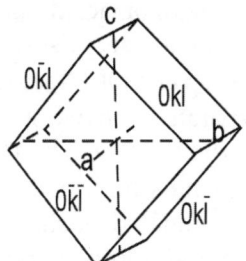

Fig. 2.94: First-order prism of orthorhombic system

v. Second-order prism (h0l)$_4$: It is an open form having four faces each of which remains parallel with b-crystallographic axis and intersects a- and c-crystallographic axes at unequal lengths. The form symbol is (h0l)$_4$ and unit form symbol is (101)$_4$. The second order prism is shown in Fig. 2.95. Its old name is macrodome because the four faces are parallel with macro (b) axis.

vi. Third-order prism (hk0)$_4$: It is an open form having four faces each of which remains parallel with c-crystallographic axis and intersects a- and b-crystallographic axes at unequal lengths. The form symbol is (hk0)$_4$. The third order prism is shown in Fig. 2.96.

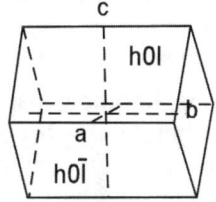

Fig. 2.95: Second-order prism of orthorhombic system

In crystallography, the 'a', 'b' and 'c' axes are in first, second and third conventional serial order. The names first-, second- and third-order refer to the crystallographic axes 'a', 'b' and 'c' respectively, to which the faces of the corresponding prism are parallel.

vii. Rhombic dipyramid (hkl)$_8$: It is a solid bounded by eight scalene triangular faces each of which intersects all the three crystallographic axes at unequal lengths. The form symbol is (hkl)$_8$. The dipyramid with symbol (hkl)$_8$ is shown in Fig. 2.97.

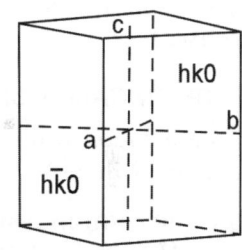

Fig. 2.96: Third-order prism of orthorhombic system

Minerals crystallizing in this class are barite, sillimanite, sulphur, topaz, olivine, stibnite, marcasite, goethite, lawsonite, andalusite, aragonite, brookite, cordierite, columbite, chrysoberyl, enstatite (and other orthopyroxenes), anthophyllite (and other orthorhombic amphiboles), etc.

2.7.2 Class mm2 (Rhombic Pyramidal Class)

Symmetry elements: The vertical crystallographic axis is the axis of 2-fold rotation. There are two vertical axial mirror planes containing a-c and b-c axis pairs. Centre of symmetry is absent.

Forms: Due to absence of the horizontal plane of symmetry, the forms at the top of the crystal are different from those at the bottom.

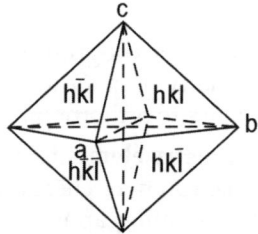

Fig. 2.97: Rhombic dipyramid of orthorhombic system

The basal pinacoid is segmented into two pedions $(001)_1$ and $(00\bar{1})_1$. The first-order prism is separated into two first-order domes, $(0kl)_2$ at the top and $(0k\bar{l})_2$ at the bottom. Similarly, the second-order prism is separated into two second-order domes, $(h0l)_2$ at the top and $(h0\bar{l})_2$ at the bottom. The rhombic dipyramid becomes two rhombic pyramids, $(hkl)_4$ at the top and $(hk\bar{l})_4$ at the bottom. The top rhombic pyramid $(hkl)_4$ is shown in Fig. 2.98.

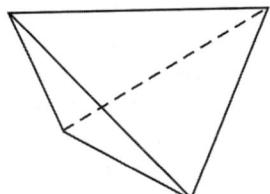

Fig. 2.98: Rhombic pyramid of orthorhombic system

Hemimorphite and bertrandite are the most common minerals crystallizing in this class.

2.7.3 Class 222 (Rhombic Disphenoidal Class)

Symmetry elements: There are three axes of 2-fold rotation coincident with the three crystallographic axes. Mirror plane and centre of symmetry are absent.

Forms: Rhombic disphenoid is the characteristic form of this class. The rhombic disphenoid is a solid bounded by four scalene triangular faces each of which intersects all the three crystallographic axes at unequal lengths. Corresponding to the rhombic dipyramid of $2/m$ $2/m$ $2/m$ class, there are two disphenoids designated as right $(hkl)_4$ (Fig. 2.99) and left $(h\bar{k}l)_4$ (Fig. 2.100), which bear enantiomorphic relationship with each other. Three pinacoids and three prisms may be present.

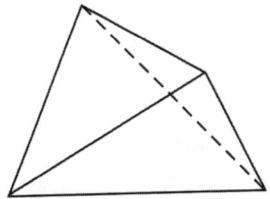

Fig. 2.99: Disphenoid (right) of orthorhombic system

Epsomite and olivenite are the common minerals crystallizing in this class.

2.8 IDENTIFICATION OF ORTHORHOMBIC CRYSTAL FORMS

Six combination forms are shown in Figs 2.101–2.106. These are to be identified with respect to form, class and system.

2.8.1 The Form Shown in Fig. 2.101 is a Combination of Four Sets of Faces, Shorter Rectangular (A), Longer Rectangular (B), Triangular (C) and Trapezoid (D) (Fig. 2.101)

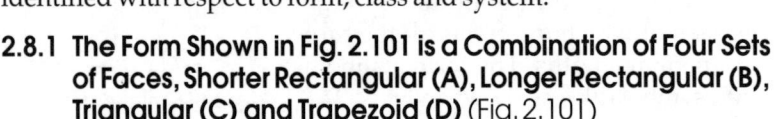

Fig. 2.100: Disphenoid (left) of orthorhombic system

Axial relationship: The a-axis joins the midpoints of opposite longer rectangular (B) faces, b-axis joins the midpoints of opposite shorter rectangular (A) faces and c-axis is perpendicular to a-b plane. The three crystallographic axes are unequal in length ($c < a < b$) but mutually perpendicular. Thus, the form represented by Fig. 2.101 belongs to orthorhombic system.

Symmetry elements: The three crystallographic axes are the axes of 2-fold rotation. There are three mirror planes each containing two of the crystallographic axes. Centre of symmetry is present. Thus, the form belongs to rhombic dipyramidal class ($2/m\,2/m\,2/m$).

Forms: There are four sets of faces in the model. The shorter rectangular (A) faces are two in number each intersects the b-axis remaining

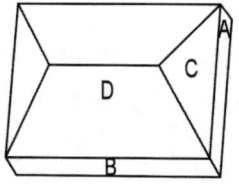

Fig. 2.101: Orthorhombic combination form

parallel with a- and c-axes. The form symbol is $(010)_2$ and it is b-pinacoid. The longer rectangular (B) faces are two in number each intersects the a-axis remaining parallel with b- and c-axes. The form symbol is $(100)_2$ and it is a-pinacoid. The triangular (C) faces are four in number (only upper side faces are shown in Fig. 2.101), each intersects the b- and c-axes at unequal lengths, remaining parallel with a-axis. The form symbol is $(0kl)_4$ and it is first-order prism. The trapezoid (D) faces are four in number (only upper side faces are shown in Fig. 2.101), each intersects the a- and c-axes at unequal lengths, remaining parallel with b-axis. The form symbol is $(h0l)_4$ and it is second-order prism.

Conclusion: The model shown in Fig. 2.101 is a combination of a-pinacoid, b-pinacoid, first-order prism and second-order prism of the dipyramidal class (2/m 2/m 2/m) of the orthorhombic system.

2.8.2 The Form Shown in Fig. 2.102 is a Combination of Three Sets of Faces, Rectangular (A), Shorter Trapezoid (B) and Longer Trapezoid (C) (Fig. 2.102)

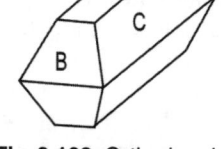

Fig. 2.102: Orthorhombic combination form

Axial relationship: The a-axis joins the midpoints of the edges between two longer trapezoid (C) faces on opposite sides, b-axis joins the midpoints of the edges between two shorter trapezoid (B) faces on opposite sides and c-axis is perpendicular to a-b plane joining the midpoints of rectangular (A) faces. The three crystallographic axes are unequal in length (c < a < b) and mutually perpendicular. Thus, the form represented by Fig. 2.102 belongs to orthorhombic system.

Symmetry elements: The three crystallographic axes are the axes of 2-fold rotation. There are three mirror planes each containing two of the crystallographic axes. Centre of symmetry is present. Thus, the form belongs to 2/m 2/m 2/m (rhombic dipyramidal) class.

Forms: There are three sets of faces in the model. The rectangular (A) faces are two in number, each intersects the c-axis remaining parallel with a- and b-axes. The form symbol is $(001)_2$ and it is basal pinacoid. The shorter trapezoid (B) faces are four in number, each intersects the b- and c-axes at unequal lengths, remaining parallel with a-axis. The form symbol is $(0kl)_4$ and it is first-order prism. The longer trapezoid (C) faces are four in number, each intersects the a- and c-axes at unequal lengths, remaining parallel with b-axis. The form symbol is $(h0l)_4$ and it is second-order prism.

Conclusion: The model shown in Fig. 2.102 is a combination of basal pinacoid, first- and second-order prisms of the rhombic dipyramidal class (2/m 2/m 2/m) of the orthorhombic system.

2.8.3 The Form Shown in Fig. 2.103 is a Combination of Four Sets of Faces, Rectangular (A), Shorter Trapezoid (B), Longer Trapezoid (C) and Quadrilateral (D) (Fig. 2.103)

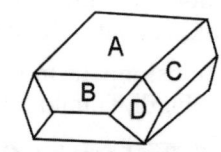

Fig. 2.103: Orthorhombic combination form

Axial relationship: The a-axis joins the midpoints of the edges between two longer trapezoid (C) faces on opposite sides, b-axis joins the midpoints of the edges between two shorter trapezoid (B) faces on opposite sides and c-axis is perpendicular to a-b plane joining the midpoints of rectangular (A) faces. The three crystallographic axes are unequal in length (c<a<b) and mutually perpendicular. Thus, the form presented in Fig. 2.103 belongs to orthorhombic system.

Symmetry elements: The three crystallographic axes are the axes of 2-fold rotation. There are three mirror planes each containing two of the crystallographic axes. Centre of symmetry is present. Thus, the form belongs to class 2/m 2/m 2/m (rhombic dipyramidal).

Forms: There are four sets of faces in the model. The rectangular (A) faces are two in number each intersects the c-axis remaining parallel with a- and b-axes. The form symbol is $(001)_2$ and it is basal pinacoid. The shorter trapezoid (B) faces are four in number, each intersects the b- and c-axes at unequal lengths, remaining parallel with a-axis. The form symbol is $(0kl)_4$ and it is first-order prism. The longer trapezoid (C) faces are four in number, each intersects the a- and c-axes at unequal lengths, remaining parallel with b-axis. The form symbol is $(h0l)_4$ and it is second-order prism. The quadrilateral (D) faces are four in number, each intersects the a- and b-axes at unequal lengths, remaining parallel with c-axis. The form symbol is $(hk0)_4$ and it is third-order prism.

Conclusion: The model shown in Fig. 2.103 is a combination of basal pinacoid, first-, second- and third-order prisms of the rhombic dipyramidal class (2/m 2/m 2/m) of the orthorhombic system.

2.8.4 The Form shown in Fig. 2.104 is a Combination of Three Sets of Faces, Smaller Trapezoid (A) Bigger Trapezoid (B) and Quadrilateral (C) (Fig. 2.104)

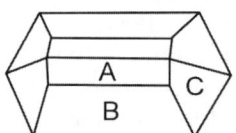

Fig. 2.104: Orthorhombic combination form

Axial relationship: The a-axis joins the midpoints of the edges between two bigger trapezoid (B) faces on opposite sides, b-axis joins the solid angles formed by the quadrilateral (C) faces in left and right and c-axis is perpendicular to a-b plane joining the midpoints of the edges between smaller trapezoid (A) faces. The three crystallographic axes are unequal in length (c<a<b) and mutually perpendicular. Thus, the form represented by Fig. 2.104 belongs to orthorhombic system.

Symmetry elements: The three crystallographic axes are the axes of 2-fold rotation. There are three mirror planes each containing two of the crystallographic axes. Centre of symmetry is present. Thus, the form belongs to (rhombic dipyramidal) 2/m 2/m 2/m class.

Forms: There are three sets of faces in the model. The smaller trapezoid (A) faces are four in number, each intersects the a- and c-axes at unequal lengths, remaining parallel with b-axis. The form symbol is $(h0l)_4$ and it is second-order prism. The bigger trapezoid (B) faces are four in number, each intersects the a- and c-axes at unequal lengths, remaining parallel with b-axis. The form symbol is $(h0l)_4$ and it is also a second-order prism. The quadrilateral (C) faces are eight in number, each intersects all the three crystallographic axes at unequal lengths. The form symbol is $(hkl)_8$ and it is rhombic dipyramid.

Conclusion: The model shown in Fig. 2.104 is a combination of two second-order prisms and rhombic dipyramid of the rhombic dipyramidal class (2/m 2/m 2/m) of the orthorhombic system.

2.8.5 The Form is a Combination of Two Sets of Faces, Trapezoid (A) and Triangular (B) (Fig. 2.105)

Axial relationship: The a-axis is front to back, b-axis is right to left and c-axis is top to bottom. The three crystallographic axes are unequal in length and mutually perpendicular. Thus, the form represented in Fig. 2.105 belongs to orthorhombic system.

Symmetry elements: The three crystallographic axes are the axes of 2-fold rotation. There are three mirror planes each containing two of the crystallographic axes. Centre of symmetry is present. Thus, the form belongs to class $2/m\,2/m\,2/m$.

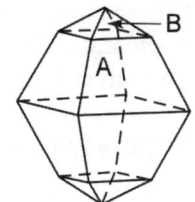

Forms: There are two sets of faces in the model. The trapezoid (A) faces are eight in number, each intersects all the crystallographic axes at unequal lengths. The form symbol is $(hkl)_8$ and it is rhombic dipyramid. The triangular (B) faces are also eight in number, each intersects all the three crystallographic axes at unequal lengths. The form symbol is $(hkl)_8$ and it is rhombic dipyramid.

Fig. 2.105: Orthorhombic combination form

Conclusion: The model shown in Fig. 2.105 is a combination of two rhombic dipyramids of the rhombic dipyramidal class $(2/m\,2/m\,2/m)$ of the orthorhombic system.

2.8.6 The Form is a Combination of Two Sets of Faces, Pentagonal (A) and Rectangular (B) (Fig. 2.106)

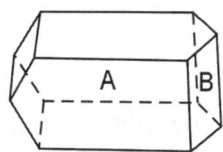

Axial relationship: The b-axis is right to left, joins the midpoints of the edges between rectangular faces; a- and c-axes are inclined to the plane of the paper. The three crystallographic axes are unequal in length (c<a<b) but mutually perpendicular. Thus, the form presented in Fig. 2.106 belongs to orthorhombic system.

Fig. 2.106: Orthorhombic combination form

Symmetry elements: The three crystallographic axes are the axes of 2-fold rotation. Plane of symmetry and centre of symmetry are absent. Thus, the form belongs to class 222 (rhombic disphenoidal).

Forms: There are two sets of faces in the model. The pentagonal (A) faces are four in number, each intersects the a- and c-axes at unequal lengths, remaining parallel with b-axis. The form symbol is $(h0l)_4$ and it is second-order prism. The rectangular (B) faces are also four in number, each intersects all the three crystallographic axes at unequal lengths. The form symbol is $(h\bar{k}l)_4$ and it is left handed rhombic disphenoid. It can be made right handed (hkl) by changing the orientation of the model.

Conclusion: The model shown in Fig. 2.106 is a combination of second-order prism and rhombic disphenoid of the rhombic disphenoidal class (222) of the orthorhombic system.

2.9 MONOCLINIC SYSTEM

Axial relationship: Crystals belonging to monoclinic system are referred to three axes of unequal length designated as a, b and c as in case of orthorhombic system. However, the a-axis is inclined to the plane containing the b and c axes. The angles between b and c axes (α) as well as a and b axes (γ) are 90°, but the angle between c and a axes (β) is different from 90°. The orientation of the crystallographic axes and the method of their notation are shown in Fig. 2.107. Previously the a-axis was called the clino-axis as it is inclined to c-axis and the b-axis the ortho-axis as it is perpendicular to c axis. In case of gypsum, a:b:c = 0.372: 1: 0.412, β = 113° 50′ and in case of orthoclase a:b:c = 0.658: 1: 0.553, β = 116° 01′.

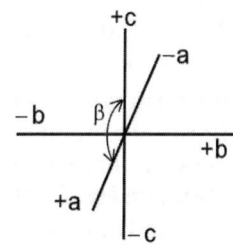

Fig. 2.107: Crystallographic axes of monoclinic system

Three classes are grouped under this system. These are 2/m (prismatic), m (domatic) and 2 (sphenoidal). Characteristic symmetry elements of each class and different forms available in them are given below.

2.9.1 Class 2/m (Prismatic Class)

The *prismatic class* is known as the *normal class of monoclinic system* as it shows maximum symmetry among all the classes belonging to this system and *gypsum type* after the characteristic mineral gypsum that crystallizes in this class.

Symmetry elements: The b-crystallographic axis is the axis of 2-fold rotation, a-c axial plane is the only mirror plane and centre of symmetry is present.

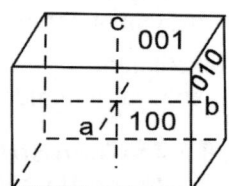

Fig. 2.108: a-, b- and c-pinacoids of monoclinic system

Forms: Pinacoids and prisms are the characteristic forms of this class.

i. **Front or a-pinacoid (100)₂:** It is an open form having two faces each of which intersects a-crystallographic axis remaining parallel with the other two crystallographic axes. The form symbol is $(100)_2$ and two faces are 100 and $\bar{1}$00 (Fig. 2.108). Its old name is orthopinacoid, as the faces are parallel with ortho (b) axis.

ii. **Side or b-pinacoid (010)₂:** It is an open form having two faces each of which intersects b-crystallographic axis remaining parallel with other two crystallographic axes. The form symbol is $(010)_2$ and two faces are 010 and $0\bar{1}0$ (Fig. 2.108). Its old name is clinopinacoid, as the faces are parallel with clino (a) axis.

iii. **Basal or c-pinacoid (001)₂:** It is an open form having two faces each of which intersects c-crystallographic axis remaining parallel with other two crystallographic axes. The form symbol is $(001)_2$ and two faces are 001 and $00\bar{1}$.

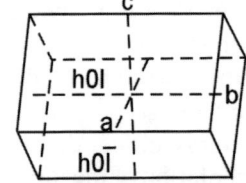

Fig. 2.109: Second-order and b-pinacoids of monoclinic system

Figure 2.108 shows the combination of a-, b- and c-pinacoids.

iv. **Second-order pinacoid:** Since the opposite ends of the a-crystallographic axis are not interchangeable, the second-order prism does not exist. Instead, two similar faces remain between 001 and 100 faces (obtuse angle side of the a-c intersection), which intersect the a- and c-crystallographic axes at unequal lengths. The symbols of these two faces are h01 and $\bar{h}0\bar{1}$ and form symbol is $(h0l)_2$. It is designated as positive second-order pinacoid. Similarly, in the acute angle side of the a-c intersection, two similar faces ($\bar{h}$01 and $h0\bar{1}$) occur with the form symbol ($\bar{h}0l)_2$, which constitute the negative second-order pinacoid. These two forms are independent of each other, and the presence of one does not necessitate the presence of another. The combination form of second-order pinacoids and b-pinacoid is shown in Fig. 2.109. The old name of second-order pinacoid is hemiorthodome, as the faces are parallel with ortho (b) axis and possess half of the number of faces of a dome.

v. **First-order prism (0kl)₄:** It is an open form having four faces each of which remains parallel with a-crystallographic axis and intersects b- and c-crystallographic axes at unequal lengths. The form symbol is $(0kl)_4$. The first-order prism in combination with a-pinacoid is shown in Fig. 2.110. Its old name is clinodome, as the faces are parallel with clino (a) axis.

vi. **Third-order prism (hk0)$_4$:** It is an open form having four faces each of which remains parallel with c-crystallographic axis and intersects a- and b-crystallographic axes at unequal lengths. The form symbol is $(hk0)_4$. The third-order prism in combination with basal pinacoid is shown in Fig. 2.111.

vii. **Fourth-order prism:** It is an open form having four faces each of which intersects all the three crystallographic axes at different lengths. There are two independent forms designated as positive with form symbol $(hkl)_4$ and negative with form symbol $(\bar{h}kl)_4$. These prisms in association with other forms are shown in Figs 1.116 and 1.117.

Many minerals crystallize in this class. Some common minerals are gypsum, azurite, chlorite, diopside, epidote, kaolinite, malachite, monazite, muscovite, orpiment, orthoclase, realgar, sphene, talc, tremolite, wolframite, etc.

2.9.2 Class m (Domatic Class)

Symmetry elements: There is only one vertical mirror plane that contains the a- and c-crystallographic axes. Axis and centre of symmetry are absent.

Forms: Dome is the characteristic form of this class. It is a two-faced form symmetrical across a mirror plane (Fig. 2.112). The first-, third- and fourth-order prisms are represented by domes $(0kl)_2$, $(0k\bar{l})_2$, $(hk0)_2$, $(\bar{h}k0)_2$, $(hkl)_2$, $(\bar{h}kl)_2$, etc. The b-pinacoid remains as a pinacoid while a-, c- and second-order pinacoids are differentiated into pedions.

The rare minerals hilgardite and clinohedrite crystallize in this class.

2.9.3 Class 2 (Sphenoidal Class)

Symmetry elements: The b-crystallographic axis is the only axis of 2-fold rotation. Plane and centre of symmetry are absent.

Forms: Sphenoid is the characteristic form of this class. A sphenoid is an open form having two faces, which are symmetrical about a 2-fold rotation axis (Fig. 2.113). The first-, third- and fourth-order prisms degenerate into pairs of enantiomorphic sphenoids like $(0kl)_2$, $(0\bar{k}l)_2$, $(hk0)_2$, $(h\bar{k}0)_2$, $(hkl)_2$, $(h\bar{k}l)_2$, etc. The a-, c- and second-order pinacoids remain as such while b-pinacoid is represented by two pedions $(010)_1$ and $(0\bar{1}0)_1$.

Pickeringite crystallizes in this class.

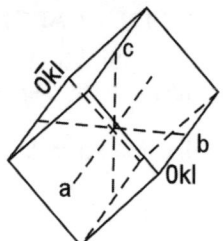

Fig. 2.110: First-order prism of monoclinic system

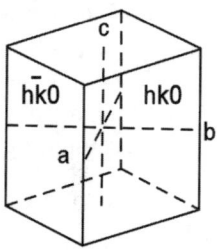

Fig. 2.111: Third-order prism of monoclinic system

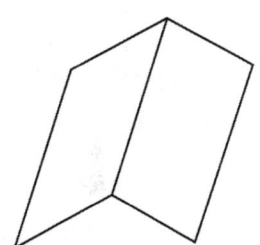

Fig. 2.112: Dome of monoclinic system

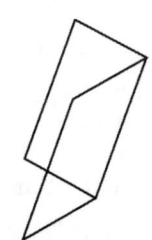

Fig. 2.113: Sphenoid of monoclinic system

2.10 IDENTIFICATION OF MONOCLINIC CRYSTAL FORMS

Eight combination forms are shown in Figs 2.114 – 2.121. These are to be identified with respect to form, class and system.

2.10.1 The Form is a Combination of Three Sets of Faces, Smaller Pentagonal (A), Triangular (B) and Bigger Pentagonal (C) (Fig. 2.114)

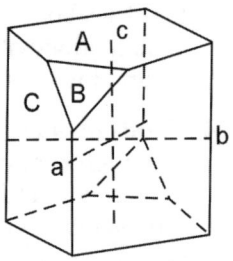

Fig. 2.114: Monoclinic combination form

Axial relationship: The three crystallographic axes are unequal in length and a-axis is inclined to the b-c axial plane, which is vertical. Thus, the form represented by Fig. 2.114 belongs to monoclinic system.

Symmetry elements: The b-axis is the axis of 2-fold rotation and a-c axial plane is the only mirror plane. Centre of symmetry is present. Thus, the form belongs to prismatic (2/m) class.

Forms: There are three sets of faces in the model. The smaller pentagonal (A) faces are two in number each intersects the c-axis remaining parallel with a- and b-axes. The form symbol is $(001)_2$ and it is basal pinacoid. The triangular (B) faces are two in number each intersects the a- and c-axes remaining parallel with b-axis. The form symbol is $(h0l)_2$ and it is positive second-order pinacoid. The bigger pentagonal (C) faces are four in number, each intersects the a- and b-axes at unequal lengths, remaining parallel with c-axis. The form symbol is $(hk0)_4$ and it is third-order prism.

Conclusion: The model shown in Fig. 2.114 is a combination of basal pinacoid, positive second-order pinacoid and third-order prism belonging to the prismatic class (2/m) of the monoclinic system.

2.10.2 The Form is a Combination of Three Sets of Faces, Smaller Pentagonal (A), Triangular (B) and Bigger Pentagonal (C). This form is Generated from the form Represented by Fig. 2.114 by Rotation of 180° About the c-crystallographic Axis (Fig. 2.115)

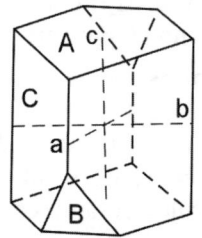

Fig. 2.115: Monoclinic combination form

Axial relationship: The three crystallographic axes are unequal in length and a-axis is inclined to the b-c axial plane, which is vertical. Thus, the form represented by Fig. 2.115 belongs to monoclinic system.

Symmetry elements: The b-axis is the axis of 2-fold rotation and a-c axial plane is the only mirror plane. Centre of symmetry is present. Thus, the form belongs to prismatic (2/m) class.

Forms: There are three sets of faces in the model. The smaller pentagonal (A) faces are two in number each intersects the c-axis remaining parallel with a- and b-axes. The form symbol is $(001)_2$ and it is basal pinacoid. The triangular (B) faces are two in number each intersects the a- and c-axes remaining parallel with b-axis. The form symbol is $(\bar{h}0l)_2$ and it is negative second-order pinacoid. The bigger pentagonal (C) faces are four in number, each intersects the a- and b-axes at unequal lengths, remaining parallel with c-axis. The form symbol is $(hk0)_4$ and it is third-order prism.

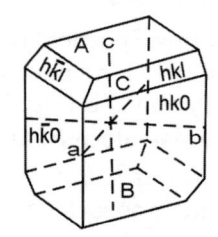

Fig. 2.116: Monoclinic combination form

Conclusion: The model shown in Fig. 2.115 is a combination of basal pinacoid, negative second-order pinacoid and third-order prism belonging to the prismatic class (2/m) of the monoclinic system.

2.10.3 The Form is a Combination of Three Sets of Faces, Rectangular (A), Pentagonal (B) and Trapezoid (C) (Fig. 2.116)

Axial relationship: The three crystallographic axes are unequal in length and a-axis is inclined to the b-c axial plane, which is vertical. Thus, the form represented by Fig. 2.116 belongs to monoclinic system.

Symmetry elements: The b-axis is the axis of 2-fold rotation and a-c axial plane is the only mirror plane. Centre of symmetry is present. Thus, the form belongs to prismatic (2/m) class.

Forms: There are three sets of faces in the model. The rectangular (A) faces are two in number each intersects the c-axis remaining parallel with a- and b-axes. The form symbol is $(001)_2$ and it is basal pinacoid. The pentagonal (B) faces are four in number, each intersects the a- and b-axes at unequal lengths, remaining parallel with c-axis. The form symbol is $(hk0)_4$ and it is third-order prism. The trapezoid (C) faces are four in number each intersects all the three crystallographic axes at unequal lengths. The form symbol is $(hkl)_4$ and it is positive fourth-order prism.

Conclusion: The model shown in Fig. 2.116 is a combination of basal pinacoid, third-order prism and positive fourth-order prism belonging to the prismatic class (2/m) of the monoclinic system.

2.10.4 The Form is a Combination of Three Sets of Faces, Rectangular (A), Pentagonal (B) and Trapezoid (C). This form is Generated from the form Represented by Fig. 2.116 by Rotation of 180° About the c-crystallographic Axis (Fig. 2.117)

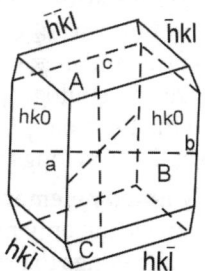

Fig. 2.117: Monoclinic combination form

Axial relationship: The three crystallographic axes are unequal in length and a-axis is inclined to the b-c axial plane, which is vertical. Thus, the form represented by Fig. 2.117 belongs to monoclinic system.

Symmetry elements: The b-axis is the axis of 2-fold rotation and a-c axial plane is the only mirror plane. Centre of symmetry is present. Thus, the form belongs to prismatic (2/m) class.

Forms: There are three sets of faces in the model. The rectangular (A) faces are two in number each intersects the c-axis remaining parallel with a- and b-axes. The form symbol is $(001)_2$ and it is basal pinacoid. The pentagonal (B) faces are four in number, each intersects the a- and b-axes at unequal lengths, remaining parallel with c-axis. The form symbol is $(hk0)_4$ and it is third-order prism. The trapezoid (C) faces are four in number each intersects all the three crystallographic axes at unequal lengths. The form symbol is $(\bar{h}kl)_4$ and it is negative fourth-order prism.

Conclusion: The model shown in Fig. 2.117 is a combination of basal pinacoid, third-order prism and negative fourth-order prism belonging to the prismatic class (2/m) of the monoclinic system.

2.10.5 The Form is a Combination of Three Sets of Faces, Rectangular (A), Bigger Trapezoid (B) and Smaller Trapezoid (C) (Fig. 2.118)

Axial relationship: The c-crystallographic axis is vertical, joins the midpoints of the edges between C-faces in top and bottom; b-axis is right to left, joins the midpoints of opposite A-faces; a-axis joins the midpoints of the edges between B-faces. The three crystallographic axes are unequal

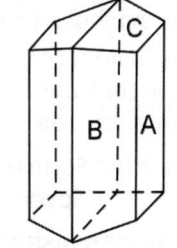

Fig. 2.118: Monoclinic combination form

in length and a-axis is inclined to the b-c axial plane, which is vertical. Thus, the form represented by Fig. 2.118 belongs to monoclinic system.

Symmetry elements: The b-axis is the axis of 2-fold rotation and a–c axial plane is the only mirror plane. Centre of symmetry is present. Thus, the form belongs to prismatic (2/m) class.

Forms: There are three sets of faces in the model. The rectangular (A) faces are two in number each intersects the b-axis remaining parallel with a- and c-axes. The form symbol is $(010)_2$ and it is b-pinacoid. The bigger trapezoid (B) faces are four in number, each intersects the a- and b-axes at unequal lengths, remaining parallel with c-axis. The form symbol is $(hk0)_4$ and it is third-order prism. The smaller trapezoid (C) faces are four in number each intersects b- and c-crystallographic axes at unequal lengths remaining parallel with a-crystallographic axis. The form symbol is $(0kl)_4$ and it is first-order prism.

Conclusion: The model shown in Fig. 2.118 is a combination of b-pinacoid, first- and third-order prisms belonging to the prismatic class (2/m) of the monoclinic system.

2.10.6 The Form is a Combination of Five Sets of Faces, Two Hexagonal (A and B), One Quadrilateral (C) and Two Pentagonal (D and E) (Fig. 2.119)

Axial relationship: The c-crystallographic axis is vertical, joins the top and bottom solid angles, b-axis is right to left, joins the midpoints of B faces and a-axis is front to back, joins the midpoints of the A-faces. The three crystallographic axes are unequal in length and a-axis is inclined to the b-c axial plane, which is vertical. Thus, the form represented by Fig. 2.119 belongs to monoclinic system.

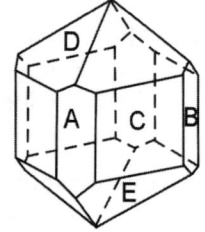

Symmetry elements: The b-axis is the axis of 2-fold rotation and a-c axial plane is the only mirror plane. Centre of symmetry is present. Thus, the form belongs to prismatic (2/m) class.

Fig. 2.119: Monoclinic combination form

Forms: There are five sets of faces in the model. The hexagonal (A) faces are two in number each intersects the a-axis remaining parallel with b- and c-axes. The form symbol is $(100)_2$ and it is a-pinacoid. The hexagonal (B) faces are two in number each intersects the b-axis remaining parallel with a- and c-axes. The form symbol is $(010)_2$ and it is b-pinacoid. The quadrilateral (C) faces are four in number, each intersects the a- and b-axes at unequal lengths, remaining parallel with c-axis. The form symbol is $(hk0)_4$ and it is third-order prism. The upper pentagonal (D) faces are four in number, each intersects all the three crystallographic axes at unequal lengths. The form symbol is $(hkl)_4$ and it is positive fourth-order prism. The lower pentagonal (E) faces are four in number, each intersects all the three crystallographic axes at unequal lengths. The form symbol is $(\bar{h}kl)_4$ and it is negative fourth-order prism.

Conclusion: The model shown in Fig. 2.119 is a combination of a-pinacoid, b-pinacoid, third-order prism, positive and negative fourth-order prisms belonging to the prismatic class (2/m) of the monoclinic system.

2.10.7 The Form is a Combination of Four Sets of Faces, Quadrilateral (A), Seven-sided (B), Triangular (C) and Trapezoid (D) (Fig. 2.120)

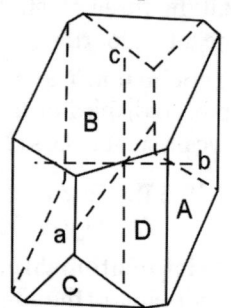

Axial relationship: The three crystallographic axes are unequal in length and a-axis is inclined to the b-c axial plane, which is vertical. Thus, the form represented by Fig. 2.120 belongs to monoclinic system.

Fig. 2.120: Monoclinic combination form

Symmetry elements: The b-axis is the axis of 2-fold rotation and a-c axial plane is the only mirror plane. Centre of symmetry is present. Thus, the form belongs to prismatic (2/m) class.

Forms: The quadrilateral (A) faces are two in number each intersects the b-axis remaining parallel with a- and c-axes. The form symbol is $(010)_2$ and it is b-pinacoid. The seven-sided (B) faces are two in number, each intersects the c-axis remaining parallel with a- and b-axes. The form symbol is $(001)_2$ and it is c-pinacoid. The triangular (C) faces are two in number, each intersects the a- and c-crystallographic axes at unequal lengths remaining parallel with b-axis. The form symbol is $(\bar{h}\,0l)_2$ and it is negative pinacoid of second order. The trapezoid (D) faces are four in number, each intersects the a- and b-axes at unequal lengths, remaining parallel with c-axis. The form symbol is $(hk0)_4$ and it is third-order prism.

Conclusion: The model shown in Fig. 2.120 is a combination of b-pinacoid, c-pinacoid, negative second order pinacoid and third-order prism belonging to the prismatic class (2/m) of the monoclinic system.

2.10.8 The Form is a Combination of Five Sets of Faces, Pentagonal (A), Hexagonal (B), Trapezoid (C), Quadrilateral (D) and Triangular (E) (Fig. 2.121)

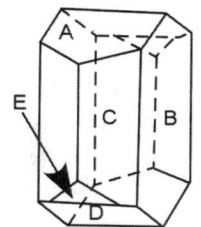

Fig. 2.121: Monoclinic combination form

Axial relationship: The c-axis is vertical joining the midpoints of the longer edges of the pentagonal (A) faces at the top and bottom. The b-axis runs from right to left joining the midpoints of hexagonal (B) faces. The a-axis is inclined to the b-c vertical plane joining the midpoints of edges between C-faces in front and back of the model. The three crystallographic axes are unequal in length and a-axis is inclined to the b-c axial plane, which is vertical. Thus, the form represented by Fig. 2.121 belongs to monoclinic system.

Symmetry elements: The b-axis is the axis of 2-fold rotation and a-c axial plane is the only mirror plane. Centre of symmetry is present. Thus, the form belongs to prismatic (2/m) class.

Forms: The pentagonal (A) faces are two in number each intersects the c-axis remaining parallel with a- and b-axes. The form symbol is $(001)_2$ and it is c-pinacoid. The hexagonal (B) faces are two in number each intersects the b-axis remaining parallel with a- and c-axes. The form symbol is $(010)_2$ and it is b-pinacoid. The trapezoid (C) faces are four in number, each intersects the a- and b-axes at unequal lengths, remaining parallel with c-axis. The form symbol is $(hk0)_4$ and it is third-order prism. The quadrilateral (D) faces are two in number, each intersects the a- and c-crystallographic axes at unequal lengths remaining parallel with b-axis. The form symbol is $(\bar{h}0l)_2$ and it is negative pinacoid of second order. The triangular (E) faces are two in number, each intersects the a- and c-crystallographic axes at unequal lengths remaining parallel with b-axis. The form symbol is $(\bar{h}\,0l)_2$ and it is negative pinacoid of second order.

Conclusion: The model shown in Fig. 2.121 is a combination of b-pinacoid, c-pinacoid, two negative pinacoids of second-order and third-order prism belonging to the class 2/m (prismatic) of the monoclinic system.

2.11 TRICLINIC SYSTEM

Crystals belonging to triclinic system are referred to three axes of unequal length, which are inclined with each other, i.e. $\alpha \neq \beta \neq \gamma \neq 90°$, where α, β and γ are the angles between b and c, c and a and a and b axes respectively. Three rules are followed in orienting a triclinic crystal and thus in determining the position of the crystallographic axes. These are: (i) The most pronounced zone should be vertical and the zone axis is the c-axis, (ii) 001 face should slope forward to the right and (iii) the directions of a and b axes are determined by the intersections of the faces 010 and 100 with 001 respectively. In proper orientation c < a < b. In old convention 'a' and 'b' axes were designated as brachy- and macro-axes respectively. The orientation of the crystallographic axes and the method of their notation are shown in Fig. 2.122. In case of axinite, a: b: c = 0.972: 1: 0.778; α = 102°41′, β = 98° 09′ and γ = 88° 08′.

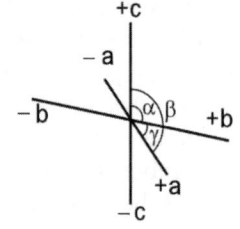

Fig. 2.122: Crystallographic axes of triclinic system

Pinacoidal ($\bar{1}$) and pedial (1) are the two classes of this system. Characteristic symmetry elements of each class and different forms available in them are given below.

2.11.1 Class $\bar{1}$ (Pinacoidal Class)

The *pinacoidal class* is known as the *normal class of triclinic system* as it shows maximum symmetry among all the classes belonging to this system and *axinite type* after the characteristic mineral axinite that crystallizes in this class.

Symmetry elements: The symmetry consists of a 1-fold axis of rotoinversion, which is equivalent to the centre of symmetry. Plane of symmetry is absent.

Forms: Pinacoids are the characteristic forms of this class.

i. **Front or a-pinacoid (100)$_2$:** It is an open form having two faces each of which intersects a-crystallographic axis remaining parallel with other two crystallographic axes. The form symbol is (100)$_2$ and two faces are 100 and $\bar{1}$ 00 (Fig. 2.123). Its old name is macropinacoid, as the faces are parallel with macro (b) axis.

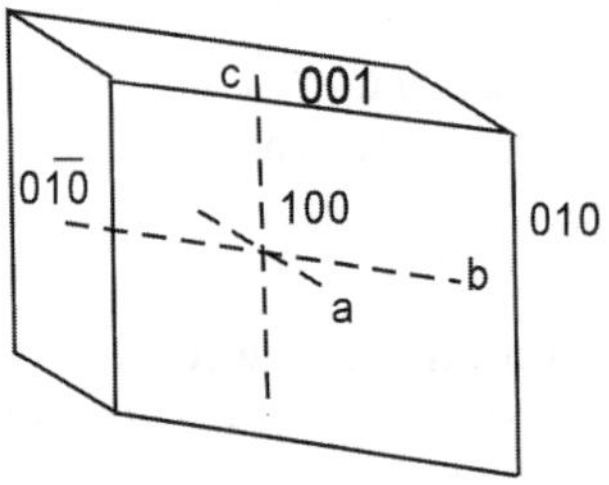

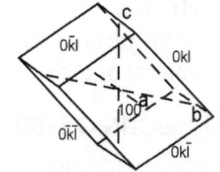

Fig. 2.123: a-, b- and c-pinacoids of triclinic system

ii. **Side or b-pinacoid (010)$_2$:** It is an open form having two faces each of which intersects b-crystallographic axis remaining parallel with other two crystallographic axes. The form symbol is (010)$_2$ and two faces are 010 and 0$\bar{1}$0 (Fig. 2.123). Its old name is brachypinacoid, since the faces are parallel with brachy (a) axis.

iii. **Basal or c-pinacoid (001)$_2$:** It is an open form having two faces each of which intersects c-crystallographic axis remaining parallel with other two crystallographic axes. The form symbol is (001)$_2$ and two faces are 001 and 00$\bar{1}$ (Fig. 2.123).

Figure 2.123 shows the combination of a-, b- and c-pinacoids.

iv. **First-order pinacoid:** It is an open form consisting of two faces each of which intersects b- and c-crystallographic axes at unequal lengths remaining parallel with a-crystallographic axis. There are two first-order pinacoids, positive with form symbol (0kl)$_2$ and negative with form symbol (0$\bar{k}$l)$_2$. The first-order pinacoids are shown in Fig. 2.124. Its old name is hemibrachydome as the two faces are parallel with brachy (a) axis.

Fig. 2.124: First-order pinacoids of triclinic system

v. **Second-order pinacoid:** It is an open form consisting of two faces each of which intersects a- and c-crystallographic axes at unequal lengths remaining parallel with b-crystallographic axis. There are two second-order pinacoids, positive with form symbol $(h0l)_2$ and negative with form symbol $(\bar{h}0l)_2$. The second-order pinacoids are shown in Fig. 2.125. Its old name is hemimacrodome as the two faces are parallel with macro (b) axis.

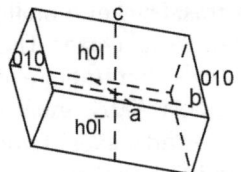

Fig. 2.125: Second-order pinacoids of triclinic system

vi. **Third-order pinacoid:** It is an open form consisting of two faces each of which intersects a- and b-crystallographic axes at unequal lengths remaining parallel with c-crystallographic axis. There are two first-order pinacoids, positive with form symbol $(hk0)_2$ and negative with form symbol $(h\bar{k}0)_2$. The third-order pinacoids are shown in Fig. 2.126.

vii. **Fourth-order pinacoid:** It is an open form consisting of two faces each of which intersects all the three crystallographic axes at unequal lengths. There are four fourth-order pinacoids, positive right $(hkl)_2$, positive left $(h\bar{k}l)_2$, negative right $(\bar{h}kl)_2$ and negative left $(\bar{h}\bar{k}l)_2$. These two-faced forms can exist independent of each other. The combination of fourth-order pinacoids is shown in Fig. 2.127.

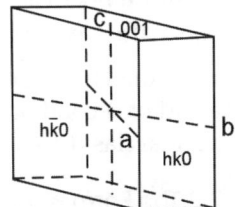

Fig. 2.126: Third-order pinacoids of triclinic system

Depending on the axial intercepts, different first-, second-, third- and fourth-order pinacoids exist.

Axinite, microcline, plagioclase feldspars, rhodonite, wollastonite, pectolite are some of the important minerals crystallizing in this class.

2.11.2 Class 1 (Pedial Class)

Symmetry elements: The symmetry element of pedial class is only an axis of 1-fold rotation, which is equivalent to no symmetry.

Forms: The pedion (Fig. 2.128) is the general form of this class, which is characterised by the presence of one face only. Various pedions are available depending on the intercept made on the crystallographic axes.

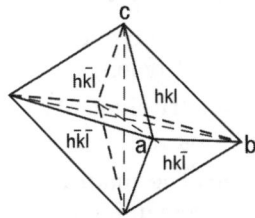

Fig. 2.127: Fourth-order pinacoids of triclinic system

2.12 IDENTIFICATION OF TRICLINIC CRYSTAL FORMS

Three combination forms are shown in Figs 2.129 – 2.131. These are to be identified with respect to form, class and system.

2.12.1 The Form is a Combination of Five Sets of Faces, one Hexagonal (A), two Pentagonal (B and C) and Two Trapezoid (D and E) (Fig. 2.129)

Axial relationship: The c-axis is vertical, b-axis runs from right to left joining the midpoints of hexagonal (A) faces and the a-axis joins the midpoints of D faces. The three crystallographic axes are unequal in length and mutually inclined with each other. Thus, the form represented by Fig. 2.129 belongs to triclinic system.

Symmetry elements: Centre of symmetry is the only symmetry element and thus, the form belongs to pinacoidal ($\bar{1}$) class.

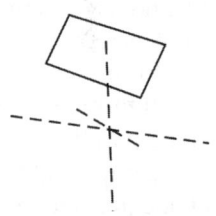

Fig. 2.128: Pedion of triclinic system

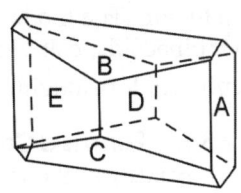

Fig. 2.129: Triclinic combination form

Forms: The hexagonal (A) faces are two in number each intersects the b-axis remaining parallel with a- and c-axes. The form symbol is $(010)_2$ and it is b-pinacoid. The pentagonal (B) faces are two in number each intersects the c-axis remaining parallel with a- and b-axes. The form symbol is $(001)_2$ and it is c-pinacoid. The pentagonal (C) faces are two in number, each intersects the a- and c-axes at unequal lengths, remaining parallel with b-axis. The form symbol is $(\bar{h}0l)_2$ and it is negative second-order pinacoid. The trapezoid (D) faces are two in number, each intersects the a- and b-crystallographic axes at unequal lengths remaining parallel with c-axis. The form symbol is $(hk0)_2$ and it is positive pinacoid of third order. The trapezoid (E) faces are two in number, each intersects the a- and b-crystallographic axes at unequal lengths remaining parallel with c-axis. The form symbol is $(h\bar{k}0)_2$ and it is negative pinacoid of third order.

Conclusion: The model shown in Fig. 2.129 is a combination of b-pinacoid, c-pinacoid, negative pinacoid of second order and positive and negative third-order pinacoids belonging to the pinacoidal class ($\bar{1}$) of the triclinic system.

2.12.2 The Form is a Combination of Six Sets of Faces, Quadrilateral (A), Two Pentagonal (B and C), Two Hexagonal (D and F) and Trapezoid (E) (Fig. 2.130)

Axial relationship: The c-axis is vertical, b-axis runs from right to left joining the midpoints of pentagonal (C) faces and the a-axis joins the midpoints of A-faces. The three crystallographic axes are unequal in length and mutually inclined with each other. Thus, the form represented by Fig. 2.130 belongs to triclinic system.

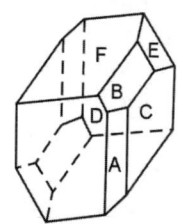

Fig. 2.130: Triclinic combination form

Symmetry elements: Centre of symmetry is the only symmetry element and thus, the form belongs to pinacoidal ($\bar{1}$) class.

Forms: The quadrilateral (A) faces are two in number, each intersects the a-axis remaining parallel with b- and c-axes. The form symbol is $(100)_2$ and it is a-pinacoid. The pentagonal (B) faces are two in number, each intersects the a- and c-axes at unequal lengths, remaining parallel with b-axis. The form symbol is $(h0l)_2$ and it is positive second-order pinacoid. The pentagonal (C) faces are two in number, each intersects the a- and b-crystallographic axes at unequal lengths remaining parallel with c-axis. The form symbol is $(hk0)_2$ and it is positive pinacoid of third order. The hexagonal (D) faces are two in number, each intersects the a- and b-crystallographic axes at unequal lengths remaining parallel with c-axis. The form symbol is $(h\bar{k}0)_2$ and it is negative pinacoid of third order. The trapezoid (E) faces are two in number, each intersects all the three crystallographic axes at unequal lengths. The form symbol is $(hkl)_2$ and it is positive right fourth order pinacoid. The hexagonal (F) faces are two in number, each intersects all the three crystallographic axes at unequal lengths. The form symbol is $(h\bar{k}l)_2$ and it is positive left fourth order pinacoid.

Conclusion: The model shown in Fig. 2.130 is a combination of a-pinacoid, positive second-order pinacoid, positive and negative third-order pinacoids and fourth order positive right and left pinacoids belonging to the pinacoidal class ($\bar{1}$) of the triclinic system.

2.12.3 The Form is a Combination of Six Sets of Faces, Three Pentagonal (A, B and F), Two Hexagonal (C and D) and One Quadrilateral (E) (Fig. 2.131)

Axial relationship: The c-axis is vertical, b-axis runs from right to left joining the midpoints of D-faces and the a-axis joins the midpoints of A-faces. The three crystallographic axes are unequal in length and mutually inclined with each other. Thus, the form represented in Fig. 2.131 belongs to triclinic system.

Symmetry elements: Centre of symmetry is the only symmetry element and thus, the form belongs to pinacoidal ($\bar{1}$) class.

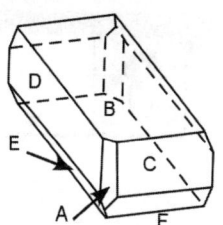

Fig. 2.131: Triclinic combination form

Forms: The pentagonal (A) faces are two in number each intersects the a-axis remaining parallel with b- and c-axes. The form symbol is $(100)_2$ and it is a-pinacoid. The pentagonal (B) faces are two in number, each intersects the c-axis remaining parallel with a- and b-axes. The form symbol is $(001)_2$ and it is c-pinacoid. The hexagonal (C) faces are two in number, each intersects the a- and b-crystallographic axes at unequal lengths remaining parallel with c-axis. The form symbol is $(hk0)_2$ and it is positive pinacoid of third order. The hexagonal (D) faces are two in number, each intersects the a- and b-crystallographic axes at unequal lengths remaining parallel with c-axis. The form symbol is $(h\bar{k}0)_2$ and it is negative pinacoid of third order. The quadrilateral (E) faces are two in number, each intersects all the three crystallographic axes at unequal lengths. The form symbol is $(\bar{h}kl)_2$ and it is negative right fourth-order pinacoid. The pentagonal (F) faces are two in number, each intersects all the three crystallographic axes at unequal lengths. The form symbol is $(\bar{h}\bar{k}l)_2$ and it is negative left fourth order pinacoid.

Conclusion: The model shown in Fig. 2.131 is a combination of a-pinacoid, c-pinacoid, positive and negative third order pinacoids and fourth order negative right and left pinacoids belonging to the pinacoidal class ($\bar{1}$) of the triclinic system.

3

Stereographic Projection

The crystals are three-dimensional polyhedral forms. The study of angular and zonal relation of the faces can be best studied by the use of various projections. The spherical projection is the fundamental one, from which other types of projections are derived. In spherical projection, the crystal is assumed to be inside a sphere, the centre of which coincides with the centre of the crystal (if the centre of symmetry is absent, the point of intersection of the crystallographic axes is taken into consideration). From the common centre, perpendiculars are drawn onto the crystal faces, which are extended till they meet the surface of the sphere. The points at which the perpendiculars touch the surface of the sphere are known as the poles of the respective faces in spherical projection. The angular relations between the crystal faces are well preserved. The faces that lie on a particular zone (tautozonal faces) plot on the same great circle (the circle whose centre is coincident with the centre of the sphere). The spherical projection, though a convenient method of crystal projection, the demerit is that it is three-dimensional by itself. The stereographic projection can be derived from the spherical projection by projecting the poles of the faces on the equatorial plane of the sphere. The projection of this plane is a circle in the stereographic projection whose diameter is equal to that of the spherical projection. This circle is designated as the primitive in stereographic projection. Imaginary lines are drawn from the poles of the spherical projection lying above and on the equatorial plane to the south pole of the sphere. The points where these imaginary lines intersect the equatorial plane are known as the poles of the crystal faces on the stereographic projection. Commonly the poles that lie in the northern hemisphere including those on the equator are transferred to the stereographic projection.

3.1 METHOD OF PLOTTING OF POLES ON THE STEREOGRAPHIC PROJECTION

A system similar to locating a place on the earth surface with the help of known latitude and longitude is adopted to locate the pole of a crystal face on stereographic projection. This is facilitated with the help of Wulff's net devised by Russian crystallographer G V Wulff. It is a stereographic net with radius of 10 cm with two sets of lines drawn at 2° intervals. A reduced version with lines drawn at 10° intervals is shown in Fig. 3.1. The lines, which join the N and S poles, are known as great circles. These are the projections of planes, which pass through the N and S poles of spherical projection with their centres coincident with reference sphere. The other types of lines, which are concentric with N and S poles, are known as small circles.

Certain facts regarding the stereographic projection are given below and shown in Fig. 3.2.

i. The poles of all crystal faces, which are parallel with the vertical crystallographic axis (100, 110, etc.) plot on the primitive (circumference).

ii. The pole of the horizontal face (001) is plotted at the centre of the stereographic projection.

iii. The angular relations between the poles of various faces are preserved in the stereographic projection but the corresponding linear distances increase from the centre towards the circumference of the projection.

iv. For all systems except hexagonal, the a and b (a_1 and a_2 for isometric and tetragonal) axes are represented by the N-S and E-W lines respectively, which are of definite length.

v. Faces parallel with a-crystallographic axes plot on the b-axis and vice versa. The face parallel with both a- and b-crystallographic axes (001) plots at the point of their intersection, i.e. centre of the stereographic projection where the vertical axis emerges. The faces which intersect all the three crystallographic axes (hkl, 111, etc.) plot within the quadrants.

vi. The tautozonal faces (faces in the same zone, i.e. whose mutual intersections are parallel with each other) plot on the same great circle. The primitive is also a great circle.

vii. The poles of certain faces are fixed. For example, the face 100, 010 and 001 are located at the positive ends of a-axis (S pole), b-axis (E pole) and centre of the stereographic projection respectively. Other faces are plotted according to their zonal and interfacial relationships.

The interfacial angle is commonly used to plot the poles of crystal faces on stereographic projection. It is defined as the angle between the perpendiculars drawn onto the concerned faces. The interfacial angles are measured by goniometers. Two types of goniometers are available. These are optical and contact types. The optical goniometer is a sophisticated instrument designed to measure the interfacial angle in case of small crystals. The contact goniometer (Fig. 3.3), on

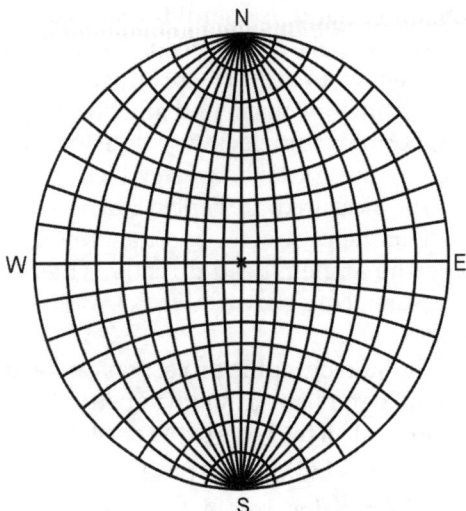

Fig. 3.1: A reduced version of the Wulff's stereographic net drawn at 10° interval

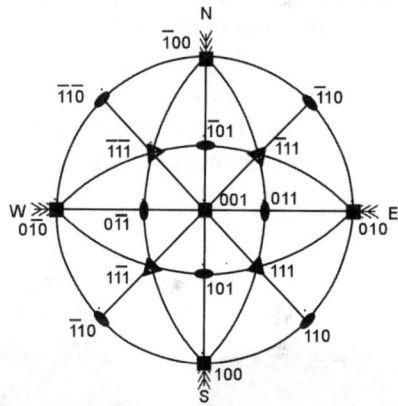

Fig. 3.2: An example of stereographic projection

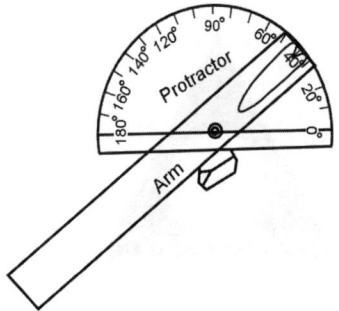

Fig. 3.3: Contact goniometer

the other hand, is useful for large crystals as well as crystal models. It consists of a semicircular protractor pivoted on an arm. The protractor is graduated from 0–180° in both clock- and anticlockwise directions (in Fig. 3.3 only anticlockwise divisions are shown). The faces of the crystal between which the interfacial angle is to be measured, are kept in tangential contact with the base of the protractor and the arm of the goniometer and the angle is read directly. The geometry of the measurement is shown in Fig. 3.4, where α is the interfacial angle.

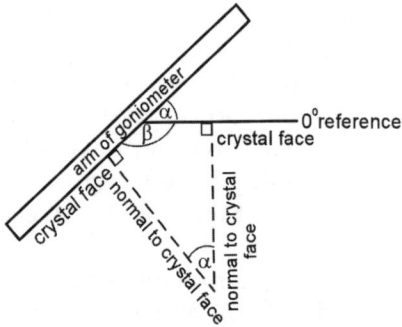

Fig. 3.4: Geometry of measurement of interfacial angle by contact goniometer

The stereographic projection can be used to exhibit crystallographic axes and symmetry elements in addition to poles of all the faces of the crystal.

i. Arrows indicate the positions of the crystallographic axes as shown in Fig. 3.2.

ii. The rotation axes of 2-, 3-, 4- and 6-fold are indicated by solid ellipse, equilateral triangle, square and regular hexagon respectively (Fig. 3.5a–d). Similarly, circle within equilateral triangle, ellipse within square and triangle within a regular hexagon indicate the rotoinversion axes of 3-, 4- and 6-fold respectively (Fig. 3.5e–g). The symbols are shown at the extremities of the axes. In Fig. 3.2, there are 3 axes of 4-fold rotation coincident with the crystallographic axes, 4 axes of 3-fold rotoinversion and 6 axes of 2-fold rotation.

iii. A solid line represents a plane of symmetry. The primitive of stereographic projection, which is generally a horizontal plane of symmetry, is indicated by a dashed line in case of crystal lacking the horizontal plane of symmetry.

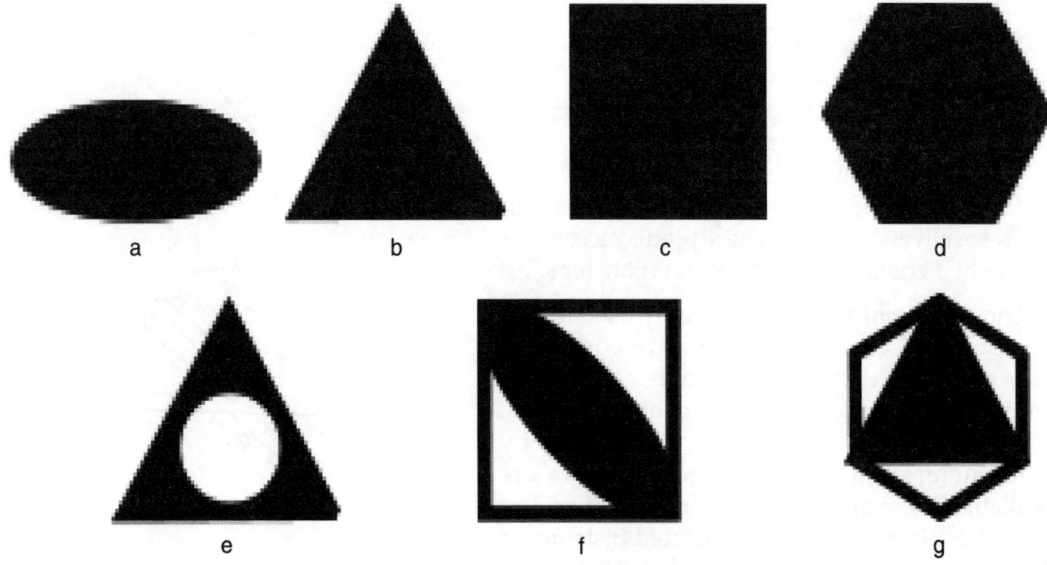

Fig. 3.5: Symbols of axes used in stereographic projection (a) 2 (b) 3 (c) 4 (d) 6 (e) $\bar{3}$ (f) $\bar{4}$ (g) $\bar{6}$

iv. The pole of a face in the upper half of the crystal (above the plane of projection) is represented by a dot (•). Some authors use cross mark (×) to represent it. The pole of a face in the lower half of the crystal (below the plane of projection) is shown by a circle. If two similar faces are present, one in upper half and other in lower half of the crystal, both of them are indicated by a dot within a circle or cross within circle.

v. The great and small circles are used for measurement of interfacial angles from the central face (001) and peripheral faces lying on the primitive respectively.

The methods of projection of some combination forms belonging to different crystal systems are shown below.

3.2 PROJECTION OF ISOMETRIC FORM

The form shown in Fig. 3.6 is a combination of cube (A), octahedron (B) and dodecahedron (C) of hexoctahedral class (4/m $\overline{3}$ 2/m) of isometric system. The procedure of projection is outlined below.

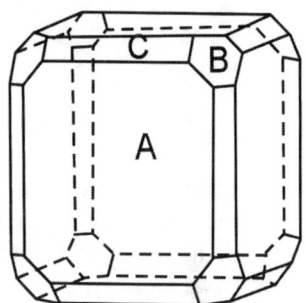

Fig. 3.6: Isometric combination form

i. Measure the interfacial angles by contact goniometer; measurements show A $\wedge$ B = 55°, A $\wedge$ C = 45° and B $\wedge$ C = 35°.

ii. Since the model has a horizontal plane of symmetry, draw the primitive by a solid line (Fig. 3.7).

iii. Locate the crystallographic axes. The N-S line is the a_1 axis, the W-E line is the a_2 axis and the a_3 axis is perpendicular at the centre of primitive (Fig. 3.7).

iv. Locate the poles of faces 001, 010, 0$\overline{1}$0, 100 and $\overline{1}$00 at their respective places. These are the poles of cube faces (A). The face 00$\overline{1}$ is located below the 001 face at the centre of the projection. Its presence is marked by encircling the pole of face 001 at the centre. Thus, all the six faces of the cube are plotted (Fig. 3.8).

v. The dodecahedral faces (C) are 12 in number; 4 are in vertical zone, i.e. parallel with the a_3-crystallographic axis and they are to be located on the primitive by dots. They are located at 45° distances from the cube faces on either side (Fig. 3.9). The face marked by C and three other faces extending in left–right direction are parallel

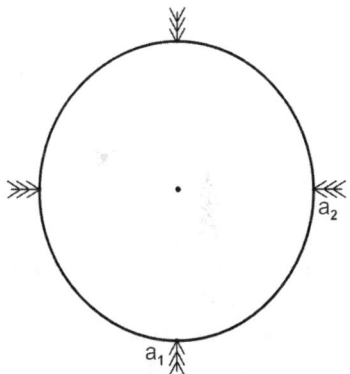

Fig. 3.7: Steps ii and iii

with a_2 crystallographic axis and thus, are to be located on the a_1 crystallographic axis at 45° distance from the cube face at the centre. Only the upper two faces are to be plotted by marking dots and the presence of lower two faces are to be indicated by encircling the poles indicated by dots. The remaining four dodecahedral faces run from front to back and parallel with the a_1 crystallographic axis. The poles of these faces will plot on the a_2 axis at 45° distances from the cube face at the centre. The upper two faces are to be plotted by marking dots and the presence of lower two faces are to be indicated by encircling the poles indicated by dots. Thus, all the 12 faces of the dodecahedron are plotted primarily on the basis of their zonal relationships (Fig. 3.9).

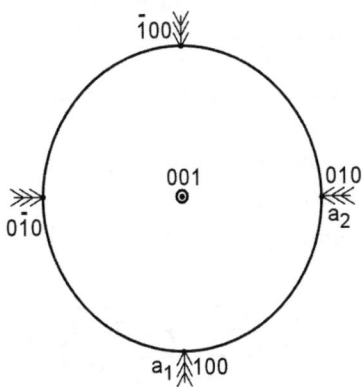

Fig. 3.8: Step iv

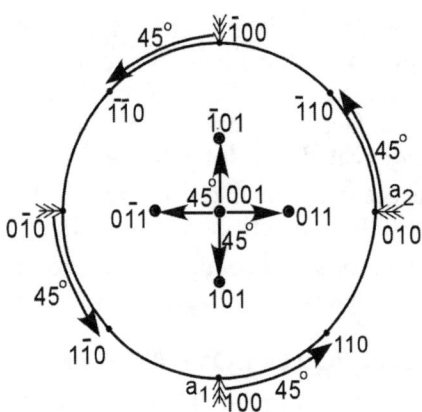

Fig. 3.9: Step v

vi. The 8 octahedron faces are not in zone with any other face. They are to be plotted by taking the interfacial angles into consideration. Two octahedral faces, 1 $\bar{1}$ 1 and 111 are located to the NW and NE of cube face 100 (Fig. 3.10) and the interfacial angles between the cube and octahedral faces are 55° in each case. So draw the 55° small circles keeping 100 face at the south pole. The interfacial angle between 111 and 110 faces is 35°. Draw the 35° small circle keeping 110 at the south pole. The point of intersection marks the pole of the 111 face. The 1 $\bar{1}$ 1 face makes an angle of 35° with 1 $\bar{1}$ 0 face. Draw the 35° small circle keeping 1 $\bar{1}$ 0 at the south pole. The point of intersection marks the pole of the 1 $\bar{1}$ 1 face (Fig. 3.10).

The poles of other two octahedral faces in the upper half of the model ($\bar{1}$ 11 and $\bar{1}$ $\bar{1}$ 1) can be plotted by drawing the 55° great circle about $\bar{1}$ 00 face and 35° small circles about $\bar{1}$ 10 and $\bar{1}$ $\bar{1}$ 0 faces respectively (Fig. 3.10). Since there are four more faces of the octahedron in the southern hemisphere, their presence are to be indicated by encircling the dots which represent four octahedral faces in the upper half of the crystal. Now the plotting of all the faces of the crystal model shown in Fig. 3.6 is over. Instead of taking a peripheral cube faces (100 or $\bar{1}$ 00), the 001 face lying at the centre of the projection can be taken into consideration while marking the octahedral faces. However, in such case the 55° great circles are to be drawn.

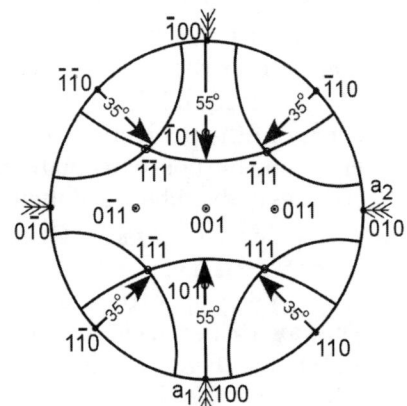

Fig. 3.10: Step vi

vii. The next step is to indicate the symmetry elements on the stereographic projection. As the model belongs to hexoctahedral class (4/m $\bar{3}$ 2/m) of the isometric system, there are 3 axes of 4-fold rotation coincident with crystallographic axes. These are to be indicated by solid squares at the extremities of the a_1 and a_2 axes as well as at the centre of the projection where a_3 axis emerges (Fig. 3.11). There are 4 axes of 3-fold rotoinversion perpendicular to octahedral faces. These are to be marked by circle within equilateral triangle at the points where the poles of the octahedral faces are located, i.e. in four quadrants of the projection (Fig. 3.11). In addition, there are 6 axes of 2-fold rotation perpendicular to the dodecahedral faces. These are to be shown by ellipses at the poles of the dodecahedral faces. There are 3 axial planes of symmetry, one horizontal, represented by the primitive and two vertical, represented by solid lines along a_1 and a_2 axis directions. In addition, there are six diagonal planes, which bisect the angles between axial planes; two of them are vertical, represented by straight lines bisecting the angle between vertical axial planes and four inclined planes represented by great circles at 45° positions between vertical (a_3) and horizontal axes (a_1 and a_2) (Fig. 3.11). The projection of the isometric crystal represented by Fig. 3.6 is now over.

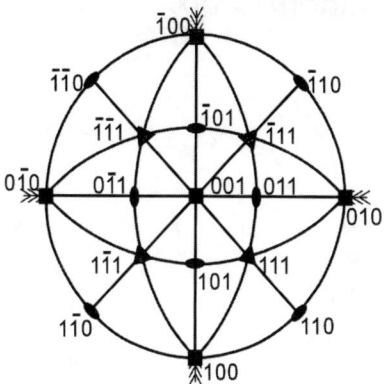

Fig. 3.11: Steps vii and viii

3.3 PROJECTION OF TETRAGONAL FORM

The form shown in Fig. 3.12 is a combination of dipyramid of first order (A), tetragonal prism of first order (B), tetragonal prism of second order (C) and basal pinacoid (D) of ditetragonal dipyramidal class (4/m 2/m 2/m) of tetragonal system. The procedure of projection is outlined below.

i. Measure the interfacial angles by contact goniometer; measurements show $A \wedge B = 40°$, $A \wedge D = 50°$, $B \wedge C = 45°$

ii. Since the model has a horizontal plane of symmetry, draw the primitive by a solid line (Fig. 3.13).

iii. Locate the crystallographic axes. The N-S line is the a_1 axis, the W-E line is the a_2 axis and the c axis is perpendicular at the centre of primitive (Fig. 3.13).

iv. Locate the poles of faces 010, $0\bar{1}0$, 100 and $\bar{1}00$ at their respective places. These are the poles of tetragonal prism of second order (C). The top face of the basal pinacoid (D) 001 is located at the centre by a dot (•) and the bottom face $00\bar{1}$ is located below the 001 face at the centre of the projection by encircling the dot (Fig. 3.14).

v. The faces of the tetragonal prism of first order (B) are parallel with the c-crystallographic axis and are in zone with the faces of second order tetragonal prism. The interfacial angle (B ∧ C) is 45°. Thus, their poles are to be located on the primitive, 45° away from the poles of the tetragonal prism of second order. The faces are 110, $\bar{1}10$, $\bar{1}\bar{1}0$ and $1\bar{1}0$ (Fig. 3.15).

vi. The faces of tetragonal dipyramid of first order (A) are in zone with the faces of basal pinacoid (D) and tetragonal prism of first order (B) and $A \wedge B = 40°$. The poles of

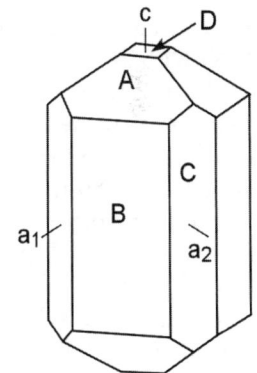

Fig. 3.12: Tetragonal combination form

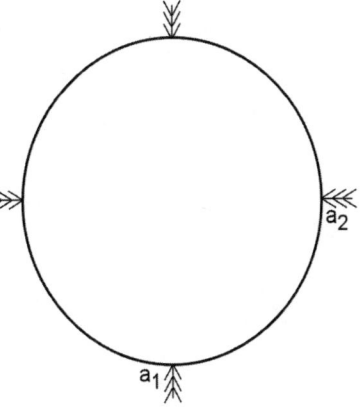

Fig. 3.13: Steps ii and iii

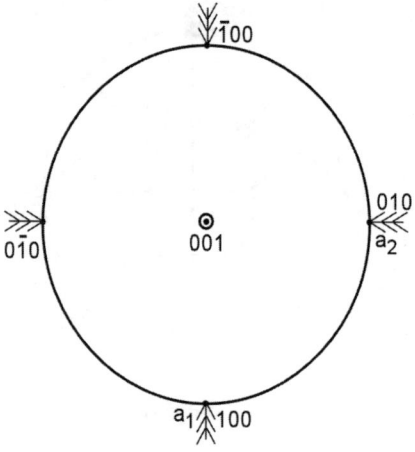

Fig. 3.14: Step iv

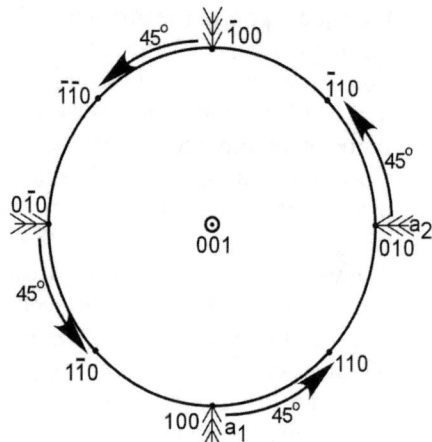

Fig.3.15: Step v

faces of tetragonal dipyramid of first order will come on the diagonal lines joining the pole of 001 face with those of tetragonal prism of first order. These are to be marked at the points 40° from the primitive on the diagonal lines (Fig. 3.16). The poles of all the 8 faces, 4 above and 4 below are indicated by dots within circles. Now the projection of all crystal faces is over.

vii. The next step is indication of symmetry elements. The vertical crystallographic axis is the axis of 4-fold rotational symmetry. This is indicated by a square at the centre of the projection. The other two crystallographic axes are the axes of 2-fold rotation, which are indicated by solid ellipses at the extremities of a_1 and a_2 axes. In addition, there are two diagonal axes of 2-fold rotation, which make 45° angles with a_1 and a_2 axes. These are marked accordingly. There are five mirror planes, one horizontal, indicated by the primitive and four vertical out of which two are coincident with horizontal crystallographic axes and the remaining two bisect the angle between vertical axial planes. These mirror planes are indicated by solid lines (Fig. 3.17). The projection of the tetragonal crystal represented by Fig. 3.12 is now over.

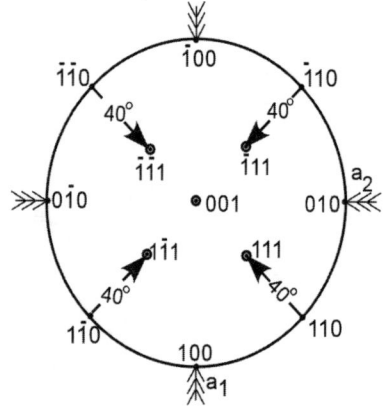

Fig. 3.16: Step vi

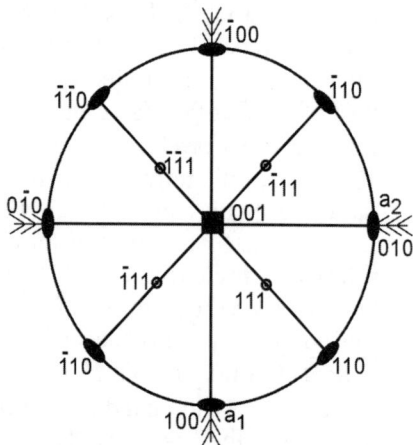

Fig. 3.17: Step vii

3.4 PROJECTION OF HEXAGONAL FORM

3.4.1 Hexagonal Division

The form shown in Fig. 3.18 is a combination of basal pinacoid (A), hexagonal dipyramid of first order (B) and hexagonal prism of first order (C) of 6/m2/m2/m class (Dihexagonal dipyramidal) of hexagonal system. The procedure of projection is outlined below.

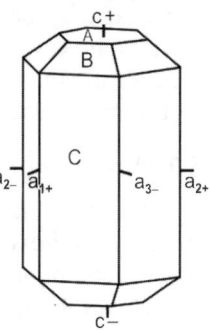

Fig. 3.18: Hexagonal combination form

i. Measure the interfacial angles by contact goniometer; measurements show $A \wedge B = 50°$ and $B \wedge C = 40°$.

ii. Since the model has a horizontal plane of symmetry, draw the primitive by a solid line (Fig. 3.19).

iii. Locate the crystallographic axes. The W-E line is the a_2 axis, positive on the right hand side; a_1 is in 120° position in the clockwise direction and a_3 is at 120° position in anticlockwise direction with respect to positive end of a_2 axis; c axis is perpendicular at the centre where three horizontal crystallographic axes intersect each other (Fig. 3.19).

iv. Two basal pinacoid faces (A) are to be plotted at the centre by dot within circle. The faces of hexagonal prism of first order (C) are located by dots on the primitive at the points exactly midway between the crystallographic axes, i.e. 30° away from the horizontal crystallographic axes (Fig. 3.19).

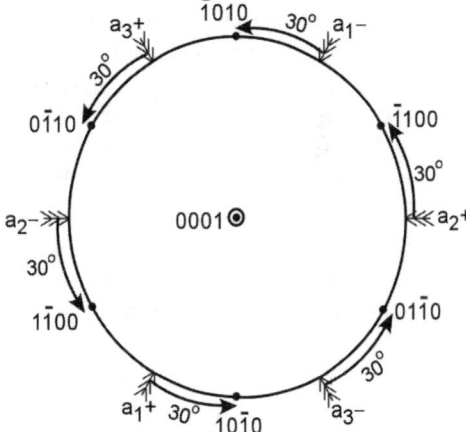

Fig. 3.19: Steps ii to iv

v. The faces of the hexagonal dipyramid of first order (B) make 40° with the faces of hexagonal prism of first order (C). Locate these faces at 40° from the faces of hexagonal prism of first order on the straight lines between the faces of hexagonal prism of first order and basal pinacoid. Since for each face in the upper part of the model there is a corresponding similar face in the lower part, the dipyramid faces are to be plotted by dot within circle (Fig. 3.20). The plotting of all the twenty faces is over.

vi. The next step is indication of symmetry elements. The vertical crystallographic axis is the axis of 6-fold rotation. This is indicated by a regular hexagon at the centre of the projection. Three horizontal crystallographic axes and three diagonal axes bisecting the angles between horizontal crystallographic axes are the axes of 2-fold rotation. These are indicated by solid ellipses at their extremities. There are seven mirror planes, one horizontal,

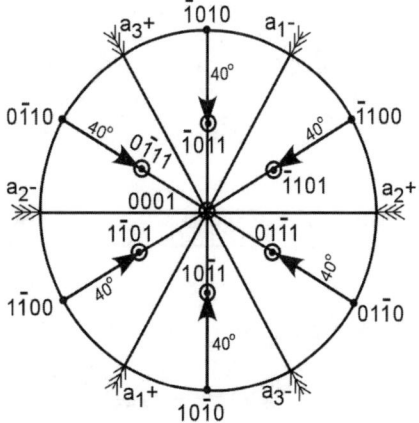

Fig. 3.20: Step v

indicated by the primitive and six vertical, coincident with the axes of 2-fold symmetry. These are shown by solid lines (Fig. 3.21). The projection of the crystal represented by Fig. 3.18 is now over.

3.4.2 Rhombohedral or Trigonal Division

The form shown in Fig. 3.22 is a combination of basal pinacoid (A), hexagonal prism of second order (B), negative rhombohedron (C) and positive scalenohedron (D) of $\bar{3}\,2/m$ (Hexagonal scalenohedral) class of hexagonal system. The procedure of projection is outlined below.

i. Measure the interfacial angles by contact goniometer; measurements show A ∧ C = 62°, B ∧ C = 38°, A ∧ D = 70° and B ∧ D = 24°

ii. Since the model has no horizontal plane of symmetry, draw the primitive by a dotted line (Fig. 3.23).

iii. Locate the crystallographic axes. The W-E line is the a_2 axis, positive at the right hand side; a_1 is in 120° position in the clockwise direction and a_3 is at 120° position in anticlockwise direction with respect to a_2 axis; c axis is perpendicular at the centre where three horizontal axes intersect (Fig. 3.23).

iv. The faces of hexagonal prism of second order (B) are located at the end of horizontal axes by dots and the basal pinacoid faces are located at the centre by dot within circle (Fig. 3.23).

v. The faces of the negative rhombohedron (C) make 62° with the faces of basal pinacoid and 38° with the prism faces. The poles of rhombohedron faces are to be marked by drawing the 62° great circles from 0001 and 38° small circles from prism faces (Fig. 3.24).

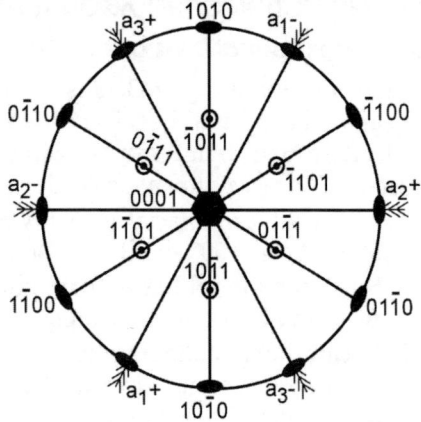

Fig. 3.21. Step vi

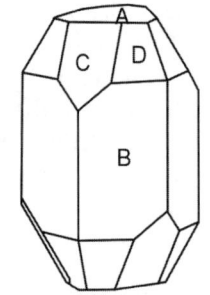

Fig. 3.22: Trigonal combination form

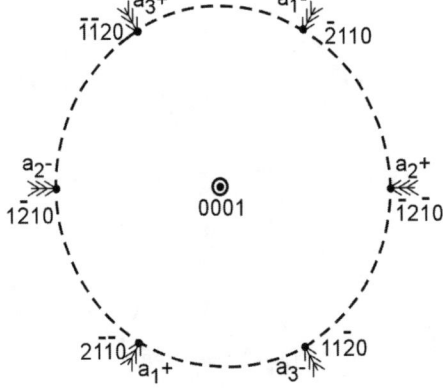

Fig. 3.23: Steps ii–iv

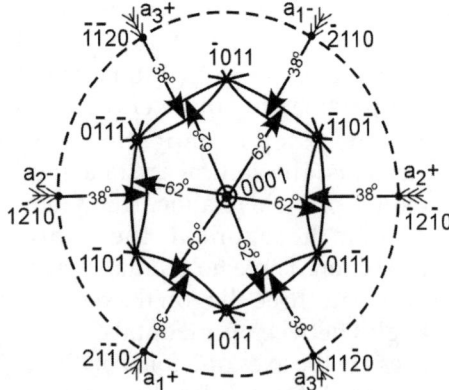

Fig. 3.24: Step v

vi. Three of the poles lie in northern hemisphere (indicated by dot) and three poles lie in southern hemisphere (indicated by circles) (Fig. 3.25).

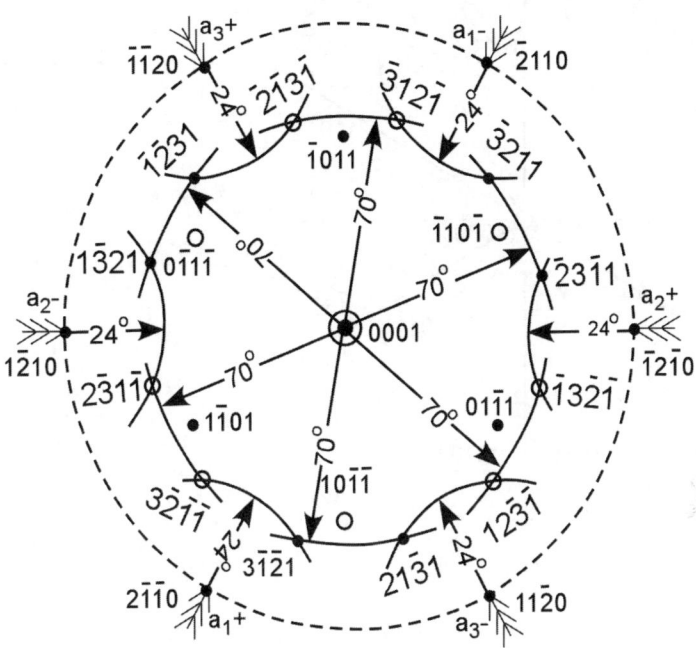

Fig. 3.25: Step vi–vii

vii. The faces of the positive scalenohedron (D) make 70° with the faces of basal pinacoid and 24° with the prism faces. The poles of scalenohedron faces are to be marked by drawing the 70° great circles from 0001 and 24° small circles from prism faces. Three pairs of the poles lie in northern hemisphere (indicated by dot) and three pairs poles lie in southern hemisphere (indicated by circles) (Fig. 3.25).

viii. The next step is indication of symmetry elements. The vertical crystallographic axis is the axis of 3-fold rotoinversion. This is indicated by a circle within a triangle at the centre of the projection. The three horizontal crystallographic axes are the axes of 2-fold rotation, which are indicated by solid ellipses at their extremities. There are three diagonal mirror planes, which bisect the angles between horizontal crystallographic axes. These are shown by solid lines (Fig. 3.26). The projection of the crystal represented by Fig. 3.22 is now over.

3.5 PROJECTION OF ORTHORHOMBIC FORM

The form shown in Fig. 3.27 is a combination of b-pinacoid (A), c-pinacoid (B), prism of first-order (C), prism of second-order (D), two prisms of third-order (120) (E), and (320) (F) and orthorhombic dipyramid (G) of 2/m 2/m 2/m (rhombic dipyramidal) class of orthorhombic system. The procedure of projection is outlined below.

i. Measure the interfacial angles by contact goniometer; measurements show $A \wedge E = 32°$, $A \wedge F = 52°$, $B \wedge C = 52°$ and $B \wedge D = 40°$, $A \wedge G = 50°$ and $B \wedge G = 57°$.

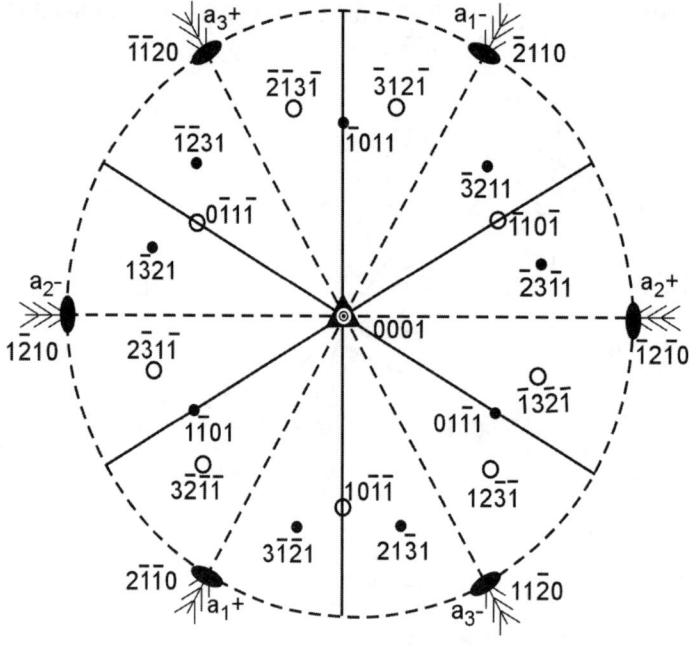

Fig. 3.26: Step viii

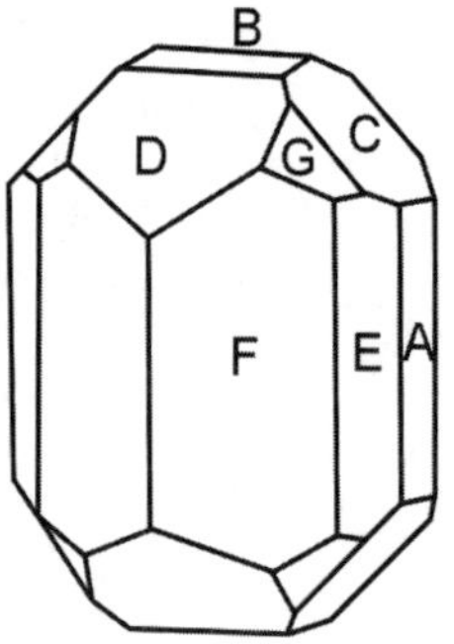

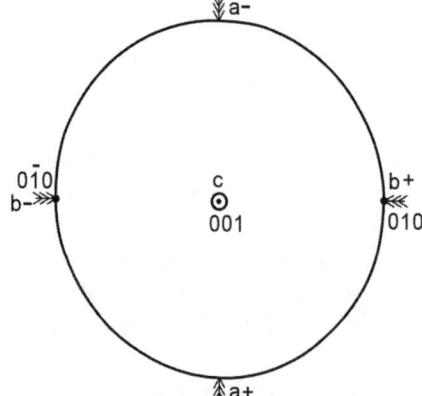

Fig. 3.27: Orthorhombic combination form

ii. Since the model has a horizontal plane of symmetry, draw the primitive by a solid line (Fig. 3.28).

iii. Locate the crystallographic axes. The N-S line is the a-axis, W-E line is the b-axis and c-axis is located at the centre of the primitive perpendicular to the plane of the paper.

iv. Locate the poles of faces 010 and 0$\bar{1}$0, at their respective places, i.e. at the right and left extremities of b-axis respectively. These are the poles of b-pinacoid (A). The top face of the basal or c-pinacoid (B) 001 is located at the centre and the bottom face, 00$\bar{1}$ is located below the face at the centre of the projection. Its presence is marked by encircling the pole of face 001 at the centre (Fig. 3.28).

v. The faces of prism of first-order (C) are in zone with the c-pinacoid faces and a-crystallographic axis is the zone axis. Thus, the poles of the prism of first-order will come on b-crystallographic axis. Since B $\wedge$ C = 52°, they are to be marked on

Fig. 3.28: Steps ii–iv

the b-crystallographic axis at distances of 52° from the centre on both sides of 001 by dot within circle. Similarly, the poles of the prism of second-order (D) are to be marked on a-crystallographic axis on both sides of 001 face at distances of 40° by dot within circle (Fig. 3.29).

vi. There are two prisms of third-order $(120)_4$ (E) and $(320)_4$ (F) which make interfacial angles of 32° and 52° with the faces of b-pinacoid respectively. All these faces lie in vertical zone and should come on the primitive. The positions of the poles are marked on the primitive at angular distances of 32° and 52° respectively from the poles of b-pinacoid on either side (Fig. 3.30).

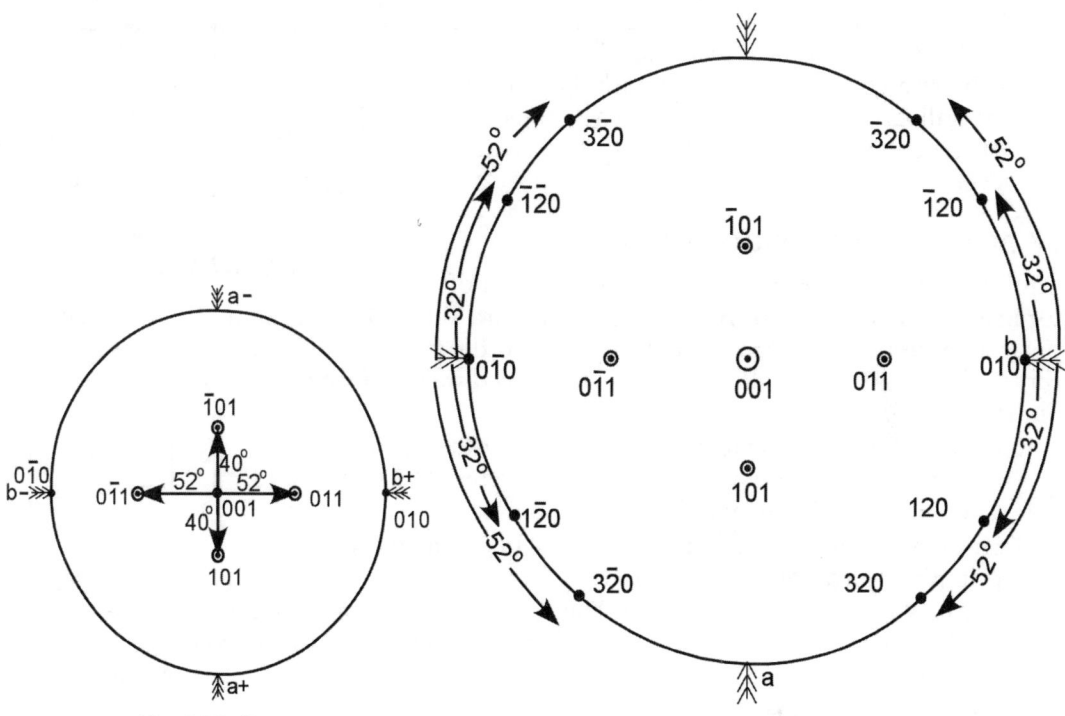

Fig. 3.29: Step v   **Fig. 3.30:** Step vi

vii. The faces of orthorhombic dipyramid make interfacial angles of 50° and 57° with the faces of b- and c-pinacoids respectively. Since the dipyramidal faces intersect all the three crystallographic axes, the poles will come within the quadrants. Their positions are to be marked by tracing 50° small circles from b-pinacoid faces and 57° great circles from the c-pinacoid face on the left and right sides. Since corresponding to each face of the dipyramid in the upper part of the model there is a corresponding face in the lower part, the faces are to be marked by dot within circle (Fig. 3.31).

viii. Now the projection of all the faces is over. The next step is to indicate the symmetry elements.

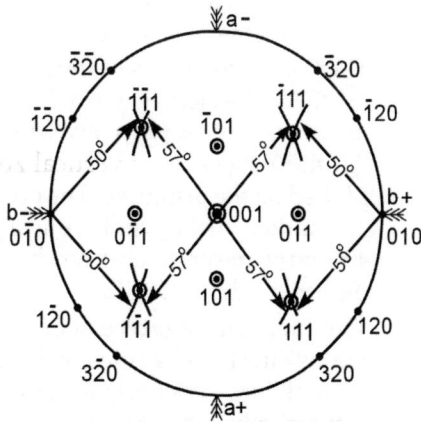

Fig. 3.31: Step vii

The three crystallographic axes are the axes of two-fold rotation. These are indicated by solid ellipses at the extremities of the crystallographic axes. There are three mirror planes, one horizontal and two vertical, each containing c- and one of the horizontal crystallographic axis. The horizontal plane has been indicated by the primitive. The vertical planes are indicated by solid straight lines coincident with the a- and b-crystallographic axes (Fig. 3.32). The projection of the crystal represented by Fig. 3.27 is now over.

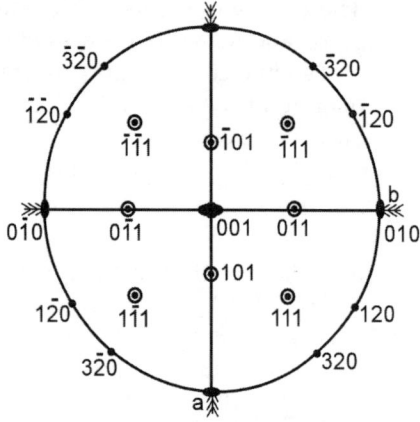

Fig. 3.32: Step viii

3.6 PROJECTION OF MONOCLINIC FORM

The form shown in Fig. 3.33 is a combination of b-pinacoid (A), c-pinacoid (B), two prisms of third-order (110) (C), and (310) (D), negative pinacoid of second-order (E) and negative third-order prism (F) of 2/m (prismatic) class of monoclinic system. The procedure of projection is outlined below.

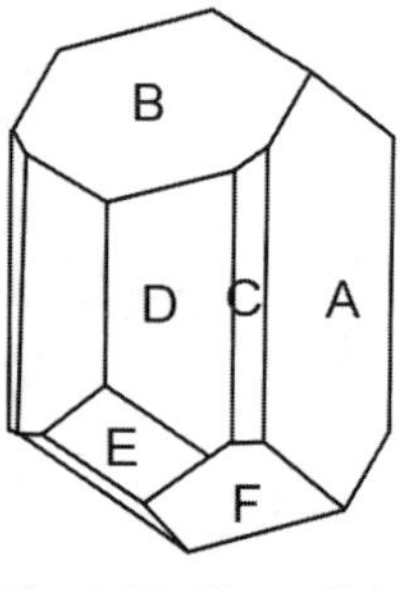

i. Measure the interfacial angles by contact goniometer; measurements show $A \wedge C = 40°$, $A \wedge D = 66°$, $D \wedge E = 30°$ and $C \wedge E = 51°$, $D \wedge F = 50°$ and $C \wedge F = 45°$.

ii. Since the model lacks a horizontal plane of symmetry, draw the primitive by a dashed line (Fig. 3.34).

iii. Locate the crystallographic axes. The W-E line is the b-axis and c-axis is located at the centre of the primitive perpendicular to the plane of the paper (Fig. 3.34).

iv. Locate the poles of faces 010, 0$\bar{1}$0, at their respective places, i.e. at the right and left extremities of b-axis respectively. These are the poles of b-pinacoid (A). The top face of the basal or c-pinacoid (B) 001 is located at the centre and the bottom face, 00$\bar{1}$ is located below the face at the centre of the projection. Its presence is indicated by encircling the pole of face 001 at the centre (Fig. 3.34).

v. There are two prisms of third-order (110)$_4$ (C) and (310)$_4$ (D) the faces of which have interfacial angles of 40° and 66° respectively with b-pinacoid faces. All these faces lie in vertical zone and are to be plotted on the primitive. The positions of the poles are marked on the primitive at angular distances of 40° and 66° respectively from the poles of b-pinacoid on either side (Fig. 3.35).

vi. The faces of negative second-order pinacoid make interfacial angles of 51° and 30° with the faces of third-order prisms (110)$_4$ (C) and (310)$_4$ (D) respectively. Since the faces are parallel with the faces of b-pinacoid, they should come on N-S line.

Fig. 3.33: Monoclinic combination form

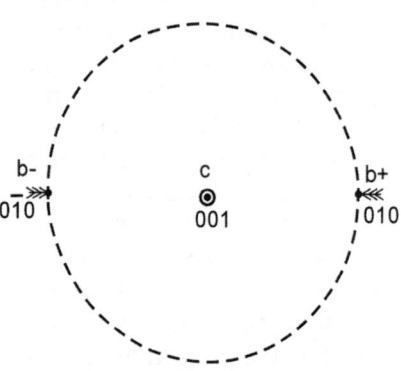

Fig. 3.34: Steps ii–iv

Their positions are to be marked by tracing 51° and 30° small circles from the faces of third-order prisms $(110)_4$ and $(310)_4$ respectively (Fig. 3.36).

vii. The faces of negative third-order prism make interfacial angles of 45° and 50° with the faces of third-order prisms $(110)_4$ (C) and $(310)_4$ (D) respectively. Since the third-order prism faces intersect all the three crystallographic axes the poles will come within the quadrants. Their positions are to be marked by tracing 45° and 50° small circles from the faces of third-order prisms $(110)_4$ and $(310)_4$ respectively (Fig. 3.37).

viii. Now the projection of all the faces is over. The next step is to indicate the symmetry elements. The b-crystallographic axis is the only axis of two-fold rotation. This is indicated by solid ellipses at the extremities of the b-crystallographic axis. There is one mirror plane containing the a- and c-crystallographic axes, which is indicated by the N-S solid line (Fig. 3.38). The projection of the crystal represented by Fig. 3.33 is now over.

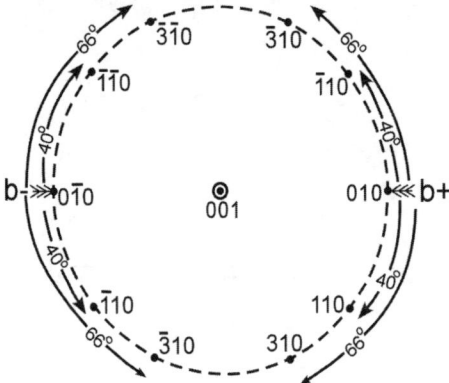

Fig. 3.35: Step v

3.7 DETERMINATION OF AXIAL RATIO

This section is devoted to determination of axial ratio from stereographic projection of the poles of crystal faces. The required data are indices of face or faces and data for locating the said face or faces on the stereogram. There are several cases in this aspect. Some of the common cases are described below.

3.7.1 Faces Lying within a Quadrant

3.7.1.1 Let the Indices of a Crystal Face are 111 and the Interfacial Angles between this Face and Faces 001 and 010 are 50° in Each Case

i. Locate the pole (P) of the face 111 by its angular relationship with faces 001 and 010 (Fig. 3.39).

ii. Join the centre O with P and extend it till it meets the primitive at K.

iii. Drop a perpendicular from B (positive end of b-axis) onto OK and extend the same till it meets the a-axis (AA'). In some cases, where

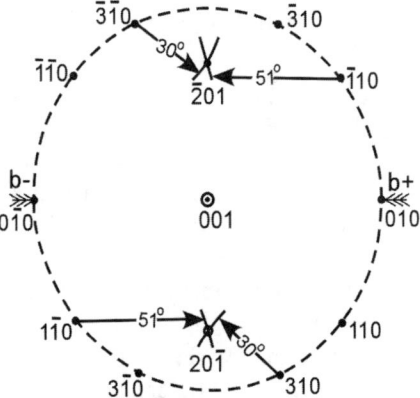

Fig. 3.36: Step vi

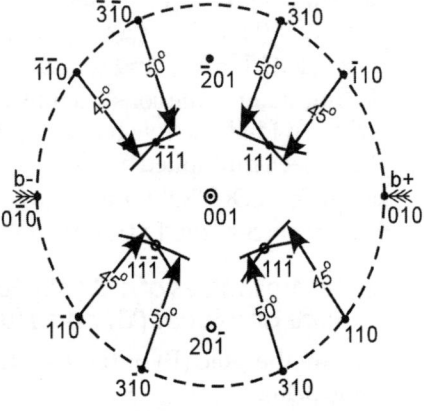

Fig. 3.37: Step vii

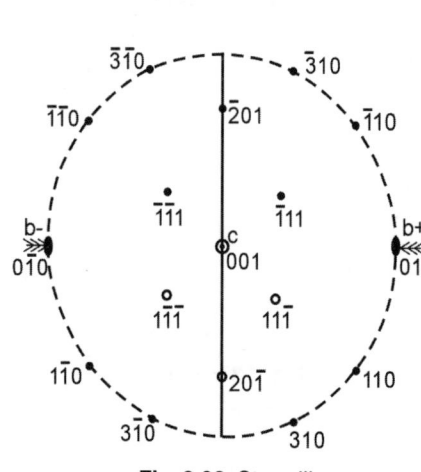

Fig. 3.38: Step viii

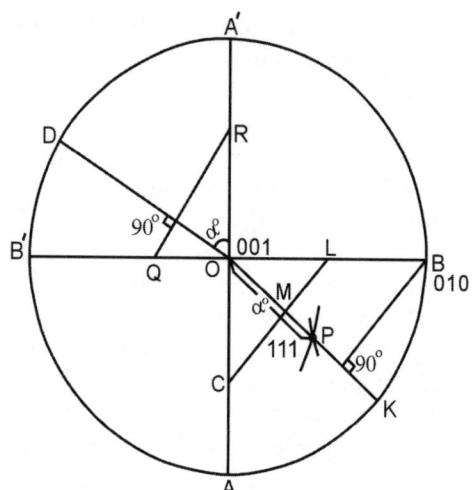

Fig. 3.39: Determination of axial ratio for the face 111

the pole of the face is located towards the primitive the point of intersection will be outside the primitive as in present case. In such cases OB is to be divided into 2 or 3 equal parts. This has to be taken into consideration while making the calculation. In the present case divide OB into 2 equal parts (OL and LB). From L draw a perpendicular onto OK, which will intersect OK and OA in the points M and C respectively. OL and OC represent half of the intercepts of the face 111 on b- and a-crystallographic axes respectively.

iv. Rotate the overlay in a clockwise direction till OK coincides with the N-S diameter of the stereographic net. Measure the angular distance OP. Let this angle is $\alpha°$.

v. Rotate the overlay to its original position. Subtend an angle equal to $\alpha°$ from OA' to the left hand side. Let it be the angle A'OD.

vi. Measure the distance OM by scale and mark the point Q on OB' so that OM = OQ.

vii. Drop a perpendicular from Q onto OD and extend it till it meets OA' at point R.

viii. **Calculations:** The indices of the face 111 suggests that it intersects the a-axis at 1a , b-axis at 1b and c-axis at 1c distances respectively.

i.e. OC = (½)a, OL = (½)b and OR = (½)c [since OB = 1b and OL = (½)OB]

⇒ a = 2OC, b = 2OL and c = 2OR

[While making the construction, the distance OB, which is equal to the intercept on b-axis was divided into two halves, i.e. OL = (½)OB. To compensate this all the three intercepts have been multiplied by 2]

⇒ a: b: c = 2OC: 2OL: 2OR

⇒ a: b: c = 2OC/2OL: 1: 2OR/2OL   [as b is commonly expressed as 1]

3.7.1.2 Let the Indices of a Crystal Face are 123 and the Interfacial Angles between this Face and Faces 001 and 010 are 60° in Each Case

i. Locate the pole (P) of the face 123 by its angular relationship with faces 001 and 010 (Fig. 3.40).

ii. Join the centre O with P and extend it till it meets the primitive at K.

iii. Since the perpendicular from B onto OK will meet the a-axis (AA') outside the primitive, divide OB into 2 equal parts (OL and LB). Drop a perpendicular LM from L onto OK and extend it till it meets OA at C. In case, the pole of the face lies in the third or fourth quadrants, perpendicular is to be drawn from B', the negative end of b-axis. OL and OC represent half of the intercepts of the face 123 on b- and a- crystallographic axes respectively.

iv. Rotate the overlay in a clockwise direction till OK coincides with the N-S diameter of the stereographic net. Measure the angular distance OP. Let this angle is α°.

v. Rotate the overlay to its original position. Subtend an angle equal to α° from OA' to the left hand side. Let it be the angle A'OD.

vi. Measure the distance OM by scale and mark the point Q on OB' so that OM = OQ.

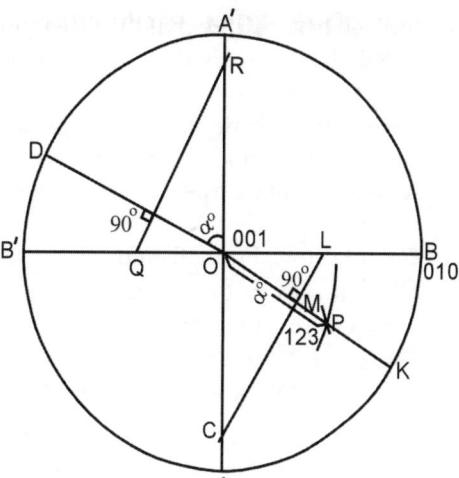

Fig. 3.40: Determination of axial ratio for the face 123

vii. Drop a perpendicular from Q onto OD and extend till it meets OA' at point R.

viii. **Calculations:** The indices of the face 123 suggests that it intersects the a-axis at 1a, b-axis at (½)b and c-axis at (1/3)c distances respectively.

i.e. OC = 1a , OL = (½)b and OR = (1/3)c

⇒ a = 2OC, b = 2 × 2OL and c = 2 × 3OR

[While making the construction, the distance OB, which is equal to the intercept on b-axis was divided into two halves, i.e. OL = (½)OB. To compensate this all the three intercepts have been multiplied by 2]

⇒ a: b: c = 2OC: 4OL: 6OR

⇒ a: b: c = 2OC/4OL: 1: 6OR/4OL

[as b is commonly expressed as 1]

3.7.2 Faces Parallel with Horizontal Axes

3.7.2.1 Let the Indices of the Crystal Faces are 101 and 011 and they make Interfacial Angles α° and β° with 001 Face Respectively

i. Locate the poles of given faces 101 and 011 on a- and b-crystallographic axes respectively with the help of the interfacial angles they make with 001. Let the poles of the faces are P and Q (Fig. 3.41).

ii. Subtend angles α° and β° to the left and right of OA' respectively. Let the angles are A'OM and A'OL.

iii. Drop a perpendicular from B' on to OM and extend it till it intersects OA' in point N.

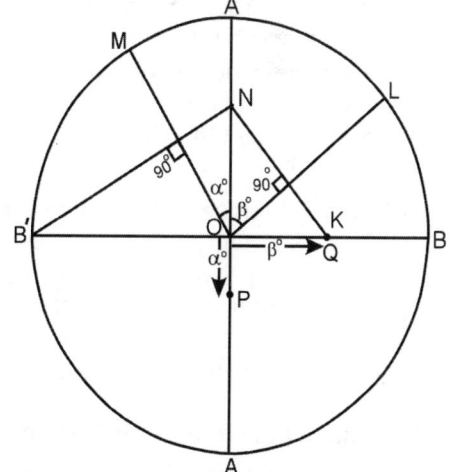

Fig. 3.41: Determination of axial ratio for the faces 101 and 011

iv. Indices 101 indicate that the face intersects a-axis at 1a distance. Had it been 201, then OB' should has been divided into two halves and the perpendicular should have been dropped from the midpoint.

v. Drop a perpendicular from N onto OL and extend till it meets OB at K. Note that ON represents 1c. Had the indices been 102, then ON should have been divided into two equal halves and the perpendicular should have been dropped from the midpoint as indicated above.

vi. **Calculations:** From the indices of the faces 101 and 011 it is evident that

OB' = 1a, ON = 1c and OK = 1b

$\Rightarrow$ a = OB', b = OK and c = ON

$\Rightarrow$ a: b: c = OB': OK: ON

or a: b: c = OB'/OK: 1: ON/OK

Had the faces been 011 and 021, then OK should had been equal to (½)b. Had the faces been 011 and 013, then ON should had been divided into three equal parts and the perpendicular onto OL should had been dropped accordingly.

3.7.2.2 Let the indices of the crystal faces are 102 and 021 and they make interfacial angles 40° and 50° respectively with the face 001

i. Locate the given faces 102 and 021 on a- and b-crystallographic axes respectively with the help of the interfacial angles they make with 001. Let the poles of the faces are P and Q (Fig. 3.42).

ii. Subtend angles 40° and 50° to the left and right of OA' respectively. Let the angles are A'OM and A'OL.

iii. Drop a perpendicular from B' on to OM and extend it till it intersects OA' in point S.

iv. Drop a perpendicular from S on to OL and extend it till it intersects OB in point K.

v. **Calculations:** Indices 102 suggests that the face intersects a- and c-axes at 1a and (½)c distances respectively; i.e. OB' = 1a and OS = (½)c $\Rightarrow$ a = OB', and c = 2OS

Indices 021 suggests that the face intersects b- and c-axes at (½)b and 1c distances respectively

vi. That is, for c = OS, b = 2OK $\Rightarrow$ for c = 2OS, b = 4OK

Hence a = OB', b = 4OK and c = 2OS $\Rightarrow$ a: b: c = OB': 4OK: 2OS

Or a: b: c = OB'/4OK: 1: 2OS/4OK

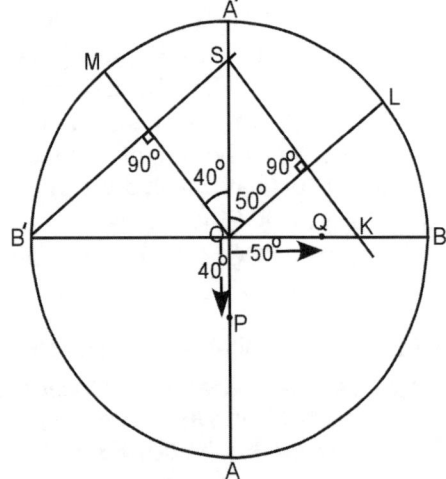

Fig. 3.42: Determination of axial ratio for the faces 102 and 021

3.7.3 One of the Faces is on the Primitive and the Other on a Horizontal Axis

i. Let the indices of the crystal face are 101 and 110. The interfacial angle between 001 and 101 is α° and the interfacial angle between 100 and 110 is β°.

ii. Locate the poles of given faces on the overlay. The pole of face 101 will come on the a-crystallographic axis (P) and the pole of face 110 will plot on the primitive (Q) (Fig. 3.43).

iii. Join O and Q and drop a perpendicular from B onto OQ. Extend BM till it intersects OA in point N.

iv. Subtend an angle equal to $\alpha°$ to the left of OA'. Let the angle is A'OL.

v. Measure the distance ON by a scale and cut off a length equal to ON from OB'. Let it be OR.

vi. From R drop a perpendicular onto OL and extend it till it intersects OA' in point S.

vii. **Calculations:** Indices 110 suggests that the face intersects a- and b-axes at 1a and 1b distances respectively, i.e. ON = 1a and OB = 1b.

Similarly the indices of the face 101 indicates that the face intersects a- and c-axes at 1a and 1c distances respectively, i.e. OR = 1a and OS = 1c.

Thus, a = ON = OR, b = OB and c = OS ⇒ a: b: c = ON: OB: OS

or a: b: c = ON/OB: 1: OS/OB.

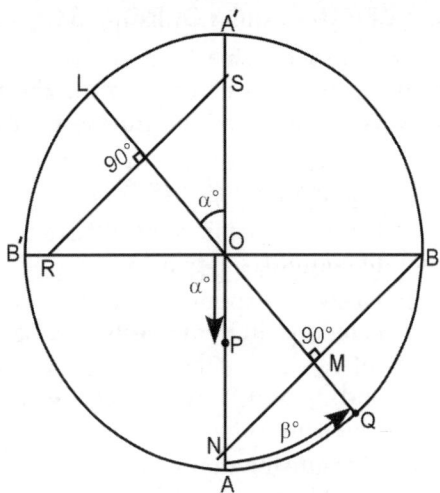

Fig. 3.43: Determination of axial ratio for the faces 101 and 001

3.8 DETERMINATION OF AXIAL RATIO (A: C) IN CASE OF HEXAGONAL SYSTEM

i. Locate the crystallographic axes. In Fig. 3.44, A_1A_1', A_2A_2' and A_3A_3' represent the a_1, a_2 and a_3 axes respectively.

ii. Locate the pole (P) of the face (say $11\bar{2}1$) from the given interfacial data (Fig. 3.44).

iii. Draw a line perpendicular to a_2 axis (A_2A_2'). Let it be ROR'.

iv. Join O and P and extend in P direction till it meets the primitive at K.

v. Join the positive end of a_1 and negative end of a_2 (A_1A_3'). Let the point of intersection of OK and A_1A_3' be Q.

vi. Measure the angular distance between O and P. Say it is $\alpha°$.

vii. Subtend an angle equal to $\alpha°$ to the left hand side of OR'. Let it be R'OS.

viii. Measure the linear distance OQ by a scale and cut off a distance equal to OQ from OA_2'. Let it be OL.

ix. Drop a perpendicular from L onto OS and extend it till it intersects OR' in point M.

x. **Calculations:** From the symbol it is evident that,

$1a = OA_1 = OA_2 = OA_3$ and $1c = OM$ ⇒
$a : c = OA_1: OM$.

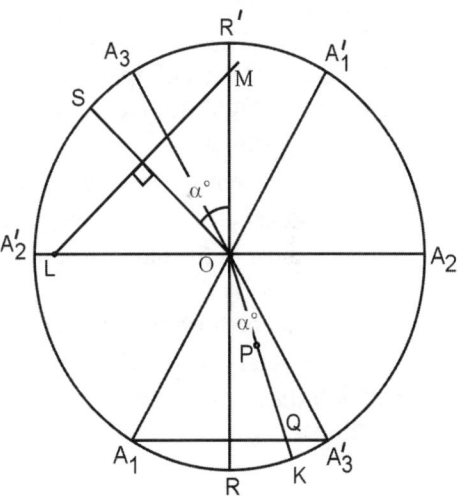

Fig. 3.44: Determination of axial ratio in hexagonal system

3.9 DETERMINATION OF INDICES OF A CRYSTAL FACE FROM STEREOGRAPHIC PROJECTION

In order to determine the indices of a face, the position of the face on the stereogram and the axial ratio are required. The detail procedure is given below.

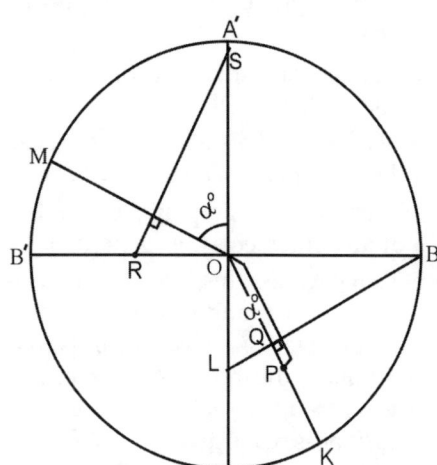

Fig. 3.45: Determination of indices of crystal face

    i. Locate the pole (P) of the crystal face on the stereogram (Fig. 3.45).

    ii. Draw a radial line OP and extend it till it meets the primitive at point K.

    iii. Draw a perpendicular from B onto OK and extend it till it intersects OA. Let the points of intersection of the perpendicular and lines OK and OA are Q and L respectively. (OB represents unit length of b-axis)

    iv. Measure the angular distance between O and P. Say it is $\alpha°$.

    v. Subtend an angle equal to $\alpha°$ from OA' to the left hand side. Let the angle is A'OM.

    vi. Measure OQ by a scale and cut off equal distance from OB'. Let it is OR.

    vii. Drop a perpendicular from R onto OM and extend it till it meets OA' at point S.

    viii. **Calculations:** OL, OB and OS represent the intercepts of the crystal face on a-, b- and c-axes respectively. These distances are to be converted into unit lengths (in terms of distances along b-axis) by dividing them by OB.

So OL/OB, OB/OB and OS/OB or say, pa, 1b and rc are the unit lengths along a-, b- and c-axes respectively.

Weiss parameters of the face are obtained by dividing these lengths by axial ratios.

In case of isometric system, where a: b: c = 1: 1: 1,

Miller indices of the face are obtained by taking the reciprocals of p, 1 and r and clearing the fractions.

In case of an orthorhombic crystal, say topaz, for which a: b: c = 0.53: 1: 0.48

The parameters are p/0.53, 1/1, r/0.48.

Miller indices are obtained by taking the reciprocal of these parameters and clearing the fractions.

4

Zone and Zonal Laws

In crystallography, a zone comprises a set of faces whose mutual intersections are parallel with each other and with a common line drawn through the centre of the crystal. The common line is known as the zone axis and the faces belonging to the same zone are known as tautozonal faces. Zonal equation refers to the mathematical relationship that exists between the faces in a zone.

A combination of cube, octahedron and dodecahedron of the hexoctahedral class is shown in Fig. 4.1. L_1 and L_2 are the intersections of the faces A (cube) and octahedron (B) and octahedron (B) and dodecahedron (C) respectively. These two lines are parallel with each other and with the common line (L_3) drawn through the centre of the crystal. Thus, the cube, octahedron and dodecahedron faces are in a zone, i.e. tautozonal and L_3 is the zone axis.

4.1 MATHEMATICAL RELATION BETWEEN INDICES OF FACES AND ZONE SYMBOL

General relationship existing between the indices of faces and zone symbol are given below.

i. If the indices of two faces lying in a zone are added, the sum is the indices of the face lying between them. In Fig. 4.1, the indices of cube face towards the observer and the dodecahedron face in the top-right position are 100 and 011 respectively. 100 + 011 = 111, which is the indices of the octahedron face that lies in between cube and dodecahedron faces.

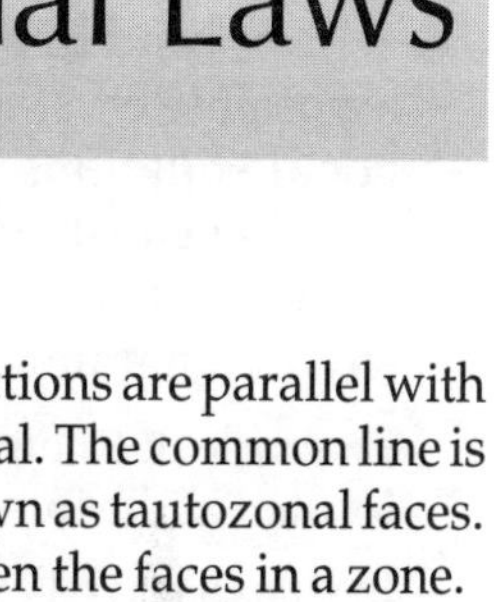

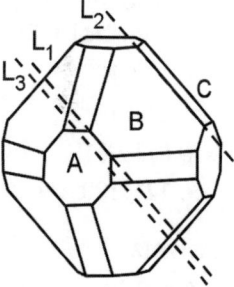

Fig. 4.1: Combination of cube, octahedron and dodecahedron of isometric system

ii. If the indices of two tautozonal faces are 'hkl' and 'pqr', the zonal symbol [uvw] is calculated as: u = a(kr – lq), v = b(lp – hr) and w = c(hq – kp), where 'a' 'b' and 'c' are unit intercepts in 'a', 'b' and 'c' axis directions. The computation scheme is shown in Fig. 4.2.

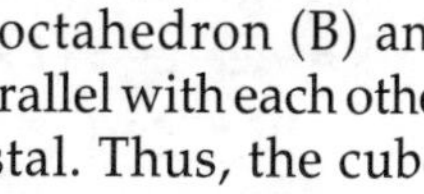

Example 1: If, 100 and 110 are two tautozonal faces of a form of the hexoctahedral class; a: b: c = 1: 1: 1.

In this case h = 1, k = 0, l = 0; p = 1, q = 1, r = 0. If [uvw] is zonal symbol

$$u = (kr - lq) = (0 \times 0) - (0 \times 1) = 0$$
$$v = (lp - hr) = (0 \times 1) - (1 \times 0) = 0$$

$$\begin{array}{c|cc|c}
h & k & l & h & k & l \\
 & \times & \times & \times & \\
p & q & r & p & q & r
\end{array}$$

Fig. 4.2: Computation of zone symbol

$w = (hq - kp) = (1 \times 1) - (0 \times 1) = 1$

∴ Zonal symbol is [001], i.e. the c-axis

iii. If a face with indices 'xyz' lies in the zone [uvw], the zonal equation $ux + vy + wz = 0$, is satisfied.

Example 2: In Fig. 4.1, it is to be checked whether the cube face intersecting the b-axis at positive end 010 lies in the zone [001] or not.

In this case $u = 0, v = 0, w = 1; x = 0, y = 1, z = 0$.

$ux + vy + wz = (0 \times 0) + (0 \times 1) + (1 \times 0) = 0 \Rightarrow$ face 010 lies in zone [001]

iv. If [uvw] and [efg] are two intersecting zones, the intersection is a possible crystal face. If the indices of the face are hkl, then the zonal equations $uh + vk + wl = 0$ and $eh + fk + gl = 0$ should be simultaneously satisfied. The values of h, k and l are calculated as: $h = vg - wf, k = we - ug$ and $l = uf - ve$. The computation scheme is shown diagrammatically in Fig. 4.3.

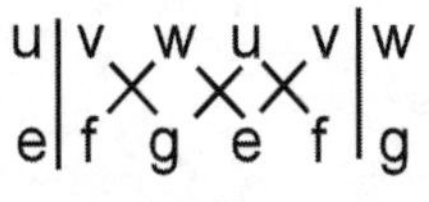

Fig. 4.3: Determination of indices of crystal face lying in two intersecting zones

Example 3: In Fig. 4.1, [001] and [010] are two intersecting zones. The symbol of the face located at the intersection of these zones is to be determined.

In this case $u = 0, v = 0, w = 1; e = 0, f = 1, g = 0$. If 'hkl' is the crystal face

$h = vg - wf = (0 \times 0) - (1 \times 1) = \overline{1}$

$k = we - ug = (1 \times 0) - (0 \times 0) = 0$

$l = uf - ve = (0 \times 1) + (0 \times 0) = 0$

∴ The symbol of the face located at the intersection of the above mentioned zones is $\overline{1}00$.

4.2 PROBLEMS RELATED TO ZONE AND ZONAL LAWS

4.2.1 The Symbols of a Pair of Tautozonal Faces are Given Below. Determine the Zonal Symbol in Each Case.

i. 120, 100	ii. 101, 001	iii. 121, 111
iv. 101, 1$\overline{1}$1	v. 111, 110	vi. 100, 101
vii. 011, 001	viii. 111, 101	ix. 1$\overline{1}$1, 1$\overline{2}$1
x. 121, 120		

Answers

i. [00$\overline{2}$]	ii. [0$\overline{1}$0]	iii. [10$\overline{1}$]	iv. [10$\overline{1}$]	v. [$\overline{1}$10]
vi. [0$\overline{1}$0]	vii. [100]	viii. [10$\overline{1}$]	ix. [10$\overline{1}$]	x. [$\overline{2}$10]

4.2.2 The Symbols of a Face and a Zone are Given Below. Determine Whether the Face Lies in the Zone or Not.

i. 230, [001]	ii. 101, [100]	iii. 121, [1$\overline{1}$1]
iv. 101, [1$\overline{1}$1]	v. 111, [010]	vi. 110, [101]
vii. 010, [001]	viii. 111, [10$\overline{1}$]	ix. 111, [11$\overline{2}$]
x. 210, [120]		

Answers

i. Yes	ii. No	iii. Yes	iv. No	v. No
vi. No	vii. Yes	viii. Yes	ix. Yes	x. No

4.2.3 The Symbols of a Pair of Zones are Given Below. In Each Case, Determine the Symbol of the Face Located at the Intersection of these Zones.

i. [012], [001]   ii. $[0\bar{1}1], [\bar{1}01]$ iii. $[121], [1\bar{1}1]$

iv. [110], [011]   v. [010], [101]   vi. [011], [001]

vii. $[011], [10\bar{1}]$ viii. $[10\bar{1}], [11\bar{2}]$

Answers

i. 100	ii. $\bar{1}\,\bar{1}\,\bar{1}$	iii. $10\bar{1}$	iv. $1\bar{1}1$	v. $10\bar{1}$
vi. 100	vii. $\bar{1}1\bar{1}$	viii. 111		

5

Physical Mineralogy

The branch of science dealing with the study of minerals is known as mineralogy. Mineral is a naturally occurring homogeneous substance with a definite (but not fixed) chemical composition and is commonly characterised by the presence of a typical internal atomic structure with or without the development of external crystalline form. Minerals are usually formed by the inorganic processes of the nature.

A hand specimen of mineral can be identified megascopically by its distinguished physical properties, which is known as *megascopic identification*. This process of identification is relatively cheap and rapid. Otherwise, a mineral may be cut into a thin slice or its surface is polished and its optical properties can be examined under a polarizing microscope. This method of identification is more accurate and is known as *microscopic identification*. Both of these methods of identification of minerals have been dealt with in this and subsequent chapters.

5.1 PHYSICAL CHARACTERS OF MINERALS

Each mineral has its own set of characteristic physical properties by which it can be identified and distinguished from other minerals. The study involves examination of mineral in naked eye or by a pocket lens (10×) and requires a hardness box, streak plate, magnet, pocket-knife and dilute hydrochloric acid, etc. The physical properties of a mineral that can be studied are its colour, lustre, streak, structure and form, fracture, hardness, tenacity, specific gravity, electrical and magnetic properties, etc.

5.1.1 Colour

Colour is a physical property that attracts the first attention. It is determined by looking at a fresh surface of a mineral in reflected light. The colour of a mineral depends on the combined effect of its composition, internal atomic arrangement, crystal structure and impurities present. These characters affect the colour absorption and reflection properties of a mineral. A mineral is green when it absorbs all the colours of the visible spectrum except green. Black and white colours are indicative of absorption of all and none of the colours of the visible light respectively. Many minerals are identified by their distinguished colour. The true colour of a mineral may be vitiated to certain extent due to the presence of impurities, play of colours caused by thin coating on the surface and by variable reflection and refraction of light caused by tarnished surfaces and cleavage cracks. A mineral is said to be colourless, when it is clear and transparent as in case of rock crystal, a variety of quartz. A particular mineral may show different colours. For example, quartz, composed of SiO_2, is commonly colourless or white, but it is also found in pink, green, brown, amethystine and even in black colours. The corundum composed of Al_2O_3, varies in colour from pale brown to deep red and dark blue. Generally minerals containing Al, Na, K, Ba and Mg as their main elements are colourless or light coloured while those with Fe, Cr, Mn, Co, Ni, Ti, V and Cu are deeply coloured. The elements, which control the colour of a mineral are called *chromophores*. Minor amounts of these elements may control the colour of a mineral, because the electrons in the d-orbit of transition metals are extremely efficient in absorbing certain visible wavelengths of light. The remaining wavelengths are reflected and give the mineral its colour. When the elements controlling the selective reflection of certain wavelengths are major components of a mineral, the mineral is called *idiochromatic* or *self-colouring*. Sphalerite is an example of idiochromatic mineral. Its colour changes from white - yellow - brown - black as its composition changes from pure ZnS to a mixture of ZnS and FeS. Ruby and sapphire are examples of *allochromatic* varieties of corundum. Small amount of Cr gives ruby deep red colour while small amount of Fe and Ti give sapphire a deep blue colour. Cr is also responsible for imparting green colour to emerald (a variety of beryl). Structural defect is also responsible for giving colour to certain minerals. For example, radiation damage imparts purple, smoky and black colours to quartz. Other causes of colouration include the oxidation or reduction of certain elements (especially Fe) and the presence of minute inclusions of other minerals. The diagnostic colours of some of the minerals are given in Table 5.1.

Table 5.1: Diagnostic colours of some minerals

Mineral	Diagnostic colour	Mineral	Diagnostic colour
Azurite	Deep blue	Kaolin	White
Chalcopyrite	Golden yellow	Malachite	Green
Covelite	Indigo blue	Orthoclase	Flesh coloured
Galena	Lead gray	Pyrite	Brass yellow
Garnet	Brown	Sulphur	Yellow
Hematite	Steel gray	Tourmaline	Black

When viewed from different angles, some minerals display a series of colours as produced by glass prism when a beam of ordinary light passes through it. Such effect is known as *play of colours*. It is shown by diamond and is produced due to dispersion of white light into its constituent colours. Some varieties of feldspars, when turned, show a series of colours like blue, green, yellow and red over broader surfaces. This phenomenon is known as *change of colours*. *Opalescence* is pearly or milky appearance exhibited by opal and moonstone (a variety of K-feldspar). *Iridescence* is the display of prismatic colours produced due to interference of light rays from incipient fractures. Minerals like quartz, calcite and mica sometimes show this effect. Some minerals show the iridescent colours due to the effect of tarnish on their surfaces by chemical reaction. Copper pyrites and erubescite (peacock ore) are good examples.

Asterism is the display of a star-shaped luminous area. It is seen in some sapphires and rubies, where it is caused by impurities of rutile. It can also occur in garnet, diopside and spinel.

Aventurescence is a reflectance effect like that of glitter. It arises from minute, preferentially oriented mineral platelets within the material. These platelets are so numerous that they also influence the mineral's body colour.

Chatoyancy refers to display luminous bands in some minerals, which appear to move as the specimen is rotated. Such minerals are composed of parallel fibers (or contain fibrous voids or inclusions), which reflect light into a direction perpendicular to their orientation, thus forming narrow bands of light. The most famous examples are tiger's eye and cymophane, but the effect may also occur in other minerals such as aquamarine, moonstone and tourmaline.

5.1.2 Diaphaneity

Diaphaneity refers to the ability of the mineral to transmit light. A mineral is said to be transparent when the outline of the object viewed through it appears sharp and distinct. Rock crystal and selenite are transparent minerals. When the object viewed through appears indistinct, the mineral is said to be subtransparent or semitransparent. A mineral, which is capable of transmitting light but cannot be seen through, is translucent. Most of the rock forming minerals belong to this category. When no light is transmitted, the mineral is said to be opaque.

5.1.3 Luminescence

Some minerals emit light when they are activated by energy other than visible light. Such effect is called luminescence. Types of luminescence are fluorescence, phosphorescence and thermoluminescence. Some minerals when exposed to certain radiations emit light instantaneously. Such effect is known as fluorescence and is best exhibited by fluorite. If the visible emission continues after the energy source is withdrawn, the effect is known as phosphorescence. Pectolite is an example of phosphorescent mineral. Thermoluminescent minerals give off visible light in response to heating. Some tourmalines, fluorite, calcite and apatite have this property.

5.1.4 Streak

The colour of the powder of a mineral is known as its streak. It may be quite different from the colour of the mineral in hand specimen. It is considered as a very distinguished character in mineral identification. The streak of a mineral is obtained by rubbing the mineral on a white unglazed porcelain plate known as streak plate. The hardness of the streak plate is 6.5–7; as such it can be used for minerals having hardness less than 7. White colour of the streak plate is produced with a scar on the plate surface when a mineral of hardness equal to or greater than seven is rubbed against the streak plate. This is to be taken into consideration while determining the streak. In such cases, the streak is determined by observing the colour of the mineral powder obtained by crushing the mineral against white background. Some minerals leave a streak similar to their natural color, such as cinnabar and lazurite. Other minerals leave different colors, such as fluorite, which always has a white streak, although it can appear in purple, blue, yellow or green colours. Hematite, which is steel gray to black in appearance, leaves a cherry red streak. Galena, which can be similar in appearance to hematite, is easily distinguished by its gray streak. The colour and streak of some minerals is given in Table 5.2.

Table 5.2: Characteristic colours and streaks of some minerals

Mineral	Colour	Streak	Mineral	Colour	Streak
Arsenopyrite	White	Black	Kaolin	White	White
Chalcopyrite	Golden yellow	Greenish black	Orthoclase	Flesh coloured	White
Covelite	Indigo blue	Black	Pyrite	Brass yellow	Brownish black
Goethite	Dark brown	Yellowish brown	Realgar	Red	Orange to yellow
Hematite	Steel gray	Cherry red	Rutile	Dark brown	Colourless
Ilmenite	Iron black	Brownish red	Sulphur	Yellow	Yellow

5.1.5 Lustre

The term lustre refers to the general appearance of a fresh mineral surface in reflected light. It is defined as the nature of shine offered by a mineral due to reflection of light from its surface. There are three main varieties of lustre, viz. metallic, submetallic and nonmetallic (Table 5.3).

Table 5.3: Lustre of minerals

Lustre	Description	Example
i. Metallic lustre	A mineral having a brilliant appearance of a metal is said to have metallic lustre. Such minerals are quite opaque to light	Galena, magnetite, chalcopyrite, hematite, etc.
ii. Submetallic lustre	The imperfect metallic lustre is termed submetallic lustre	Chromite, cuprite, psilomelane, columbite, wolframite, etc.
iii. Nonmetallic lustre	Minerals with nonmetallic lustre are generally light coloured and transmit light through thin edges	Given in Table 5.4

The following terms given in Table 5.4 are used to describe the lustre of nonmetallic minerals.

Table 5.4: Non-metallic lustres of minerals

Lustre	Description	Example
i. Adamantine	It is the most brilliant lustre shown by minerals of high refractive indices	Diamond, cerussite, sphalerite, anglesite, etc.
ii. Vitreous	The lustre of broken glass	Quartz, tourmaline, etc.
iii. Subvitreous	Imperfect vitreous lustre is termed subvitreous	Feldspar, calcite, etc.
iv. Pearly	The lustre of pearl	Talc, muscovite, brucite, selenite, apophyllite, etc.
v. Resinous	The lustre of resin	Sphalerite, sulphur, etc.
vi. Greasy	The lustre of oily glass	Nepheline, massive quartz, opal, etc.
vii. Silky	The lustre of silk	Gypsum, malachite, asbestos, etc.
viii. Earthy	The lustre of clay	Kaolinite
ix. Dull	When no lustre is visible, the lustre is said to be dull	Kaolin
	According to the amount of light reflected from the mineral surface, the following lustres have been recognised.	
x. Splendent	The surface of the mineral is sufficiently brilliant to reflect objects distinctly like those from a mirror	Galena
xi. Shining	When the surface is less brilliant and objects are reflected indistinctly, the lustre is termed shining	Pyrite, hematite, etc.
xii. Glistening	When the mineral surface is less brilliant and is incapable of giving any image	
xiii. Glimmering	Lustre feebler than glistening	

5.1.6 Structure and Form

Commonly a mineral is characterised by a definite internal atomic structure, which is manifested by the aggregation of imperfectly formed crystal grains without development of well-defined faces. In such case the mineral is said to be 'crystalline'. Under favourable physico-chemical conditions, the outer form with well developed crystal faces develop. In this case the mineral is said to be *crystallised*. Rock crystal, garnet, staurolite, etc. very often show crystallized form. A mineral is said to be *cryptocrystalline* when the degree of crystallization is observable under high power microscope. The term *amorphous* is used to describe complete lack of crystalline character. The term *massive* may be used when the mineral lacks the outer geometric form. Depending on the development of outward form and structure, certain terms are used, which have their customary meaning. These terms are explained in Table 5.5.

Table 5.5: Structure and form of minerals

Terminology	Description	Example
i. Acicular	Fine needle like crystals	Sillimanite, natrolite
ii. Amygdaloidal	Almond shaped	Zeolites
iii. Bladed	Knife-blade or lath shaped	Kyanite
iv. Blocky	Brick shaped	Orthoclase
v. Botryoidal	Small spheroidal aggregations as a bunch of grapes	Chalcedony, Psilomelane
vi. Capillary	Exhibiting a fine hair like form	Millerite, pyrite
vii. Columnar	Form resembling slender columns	Beryl, hornblende
viii. Colloform	Combination of botryoidal, globular, mamillary and reniform	Limonite, malachite
ix. Concentric	Onion shaped banding	Hematite
x. Concretionary	Spherical, ellipsoidal or irregular tuberose forms on the surface	Goethite, flint
xi. Dendritic	Branching tree or moss like form	Copper, Manganese oxide
xii. Drusy	Covered with minute implanted crystals	Quartz
xiii. Earthy	Aggregation of very minute particles producing a uniform mass	Kaolin
xiv. Equidimensional	Equally developed in all directions	Garnet
xv. Fibrous	Consisting of fine thread like strands	Asbestos, satin-spar
xvi. Foliated	Thin lamellae or leaves	Muscovite, biotite
xvii. Granular	Coarse or fine grain like	Marble
xviii. Lamellar	Separable plates or leaves like	Gypsum, wollastonite
xix. Lenticular	Flattened ball or pellet like	Azurite
xx. Mammillary	Large mutually interfering spheroidal masses/surfaces	Malachite
xxi. Massive	Compact aggregate without form	Kaolin
xxii. Oolitic	Aggregation of small spheres	Hematite, Oolitic-limestone
xxiii. Pisolitic	Aggregation of small rounded masses	Bauxite

(Contd...)

Table 5.5: Structure and form of minerals *(Contd...)*

Terminology	Description	Example
xxiv. Radiating	Crystals or fibers arranged around a central point	Stibnite
xxv. Reniform	Kidney shaped	Kidney iron-ore (hematite)
xxvi. Reticulate	Form of cross meshes like a net	Some rutiles
xxvii. Scaly	Small plate like	Tridymite
xxviii. Stalactite	Cylindrical or conical form	Calcite, aragonite
xxix. Stellate	Fibers radiating from a center to give rise to star-like forms	Wavellite
xxx. Stout	Pyramid shaped	Pyroxene
xxxi. Tabular	Like broad flat surface	Tabular spar (wollastonite)
xxxii. Tuberose	Irregular rounded surfaces	Flos-feri (aragonite)
xxxiii. Wiry (filiform)	Resembling thin wires twisted like the strands of a rope	Native silver and copper

5.1.7 Cleavage

Many of the crystallized minerals have a tendency to break along certain planes parallel to each other. These planes are known as cleavage planes and this property of the mineral is called *cleavage*. This property is related to the internal atomic structure of the mineral. In cleavable minerals, the directions of the cleavage planes are parallel to certain faces. In the plane of cleavage, the constituent atoms are more closely packed together and as such the binding force is greater in this direction than the direction at right angle to it. The cleavage plane is a plane of least cohesion, as a result of which, splitting and cleavage occurs along it. Mineral like quartz lacks cleavage, whereas others show several cleavages. Depending on the degree of their development, the cleavages are described as perfect, good, distinct, poor, indistinct (imperfect), etc. The cleavage planes commonly occur in sets. Minerals like muscovite, graphite and stibnite show one set of cleavage; feldspars and kyanite show two sets of perfect cleavages; calcite and galena have three sets of cleavages; fluorite is characterised by four sets of cleavages and sphalerite has six sets of cleavages. Due to the presence of perfect cleavages, the micas often separate in form of thin sheets.

5.1.8 Fracture

Fracture is the nature of broken surface in any direction other than the cleavage plane and is not connected with the crystalline structure of the minerals. Different types of fractures are given in Table 5.6.

Table 5.6: Different types of fractures seen in minerals

Fracture	Description
i. Conchoidal	In this case the mineral breaks with a curved concave or convex surface as in case of quartz, opal, flint, psilomelane, etc. The term subconchoidal is used for less developed conchoidal fracture.
ii. Even	The fracture surface is flat as in case of chert.
iii. Uneven	The fracture surface is rough due to the presence of minute elevations and depressions as in case of feldspars. It is the most common variety of fracture seen in case of majority of the minerals.
iv. Hackly	The fracture surface is studded with sharp and jagged elevations as in case of sillimanite.

5.1.9 Hardness

The hardness is an important property for identification of minerals in hand specimens. It is a measure of the resistance to abrasion or its ability to abrade others. Normally the hardness of a mineral is compared with a standard scale known as Mohs scale of hardness given in Table 5.7.

Table 5.7: Mohs scale of hardness

Hardness	Standard mineral	Hardness	Standard mineral
1	Talc	6	Orthoclase
2	Gypsum	7	Quartz
3	Calcite	8	Topaz
4	Fluorite	9	Corundum
5	Apatite	10	Diamond

The hardness test is made by scratching the mineral under examination with the standard minerals given in the Mohs scale of hardness. For example, if the mineral whose hardness is to be determined scratches orthoclase, then its hardness is greater than 6 but if it is scratched by quartz, then its hardness is less than 7. In such case it may be presumed that the hardness of the mineral under examination is about 6.5. Hardness may also be tested by means of fingernail, brass plate (copper coin), iron plate, glass plate and streak plate whose hardness values are 2.5, 3.5, 4.5, 5.5 and 6.5 respectively. Several precautions are to be observed while testing the hardness of a mineral. While rubbing against each other a definite scratch must be produced on the softer mineral, which can be observed with a lens after blowing away the mineral powder. The colour of the powder is the streak of the softer mineral. The hardness of a mineral is related to its atomic structure. It increases with the density of packing of ions in crystal structure, valency of the ions and with the decrease in ionic size. Since the atomic structure varies in different directions of the crystal, the hardness values also change in different directions. However, this difference is usually very small and not noticeable in case of many minerals. Kyanite is an exception. It shows two strikingly different hardness values, 5 in the direction parallel to the length of crystals and 7 in the direction perpendicular to the length of crystals. Similarly the hardness of calcite is 3 on all other faces except the basal pinacoid, where it is 2.

5.1.10 Specific Gravity

The specific gravity of a mineral is the ratio of the weight of the mineral to that of an equal volume of water. In selecting a mineral for determination of its specific gravity, the sample should be pure and free from alteration products, inclusion, etc. The specific gravity of mineral depends on the atomic weight of the constituent elements and the manner in which the atoms are packed in the crystal structure. For example, three sulphate phases with similar crystal structure, celestine ($SrSO_4$), barite ($BaSO_4$) and anglesite ($PbSO_4$) have molecular weights of 183.696, 233.426 and 303.276 respectively. Their specific gravity values are 2.9, 4.5 and 6.3 respectively. The influence of the style of packing on the specific gravity is well exhibited from two carbon minerals, graphite and diamond. Graphite with loose packing has specific gravity 2.3 while diamond with close packing has specific gravity of 2.54. In general, the nonmetallic minerals have specific gravity values of about 2.6 – 2.8 and the metallic minerals have specific gravity values of about 5 and more. However, there are some exceptions. The specific gravity of the minerals can be accurately determined by chemical balance, Walker's steelyard balance, Jolly's spring balance, pycnometer or specific gravity bottle and by using heavy liquids. However, by experience the minerals can be put in any of these specific gravity categories, viz. low (up to 2.5), moderate (2.5 – 3.5), high (3.5 – 7) and very high (7 and above).

5.1.10.1 Determination of Specific Gravity with Chemical Balance

By this method the weight of the mineral is taken in air (W_a). Then the mineral is suspended by a very fine wire and immersed in water. The weight of the mineral immersed in water (W_w) is measured. The specific gravity is calculated by the formula:

$$\text{Specific gravity} = \frac{W_a}{(W_a - W_W)}$$

5.1.10.2 Determination of Specific Gravity by Walker's Steel Yard Balance

The essential parts of the balance are the long graduated beam, which is pivoted near one end and counterbalanced by a heavy weight suspended from the short arm (Fig. 5.1). The mineral whose specific gravity is to be determined is suspended from the long arm with the help of a fine thread. Its position is adjusted by moving it along the graduated beam so that it counterbalances the constant weight. The level position of the beam is observed by a mark on the upright scale shown on the right end of the figure. The initial reading (a) of the beam with the suspended mineral sample is taken. The specimen is then immersed in water and moved along the beam until the constant weight is again balanced and a second reading (b) of the beam is taken. The readings (a) and (b) are inversely proportional to the weights of the body in air and in water respectively.

$$\text{Specific gravity} = \frac{\dfrac{1}{a}}{\dfrac{1}{a} - \dfrac{1}{b}} = \frac{b}{(b-a)}$$

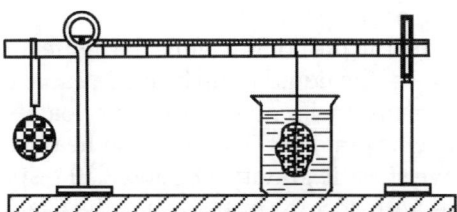

Fig. 5.1: Walker's steel yard balance

5.1.10.3 Determination of Specific Gravity by Jolly's Spring Balance

This instrument consists of a flexible spring suspended vertically against a graduated scale (Fig. 5.2). Two scale pans are attached to the lower end of the spring one above the other. The lower pan is always immersed in water. The instrument is provided with a vertical graduated mirror. The initial reading (a) of the bottom of the spring on the vertical scale is taken. A small fragment of the mineral whose specific gravity is to be determined is placed on the upper pan and a second reading (b) of the bottom of the spring on the vertical scale is taken. The mineral is then transferred to the lower pan and a third reading (c) is taken. (b–a) is proportional to the weight of the mineral in air and (b–c) is proportional to the loss of weight in water.

$$\text{Specific gravity} = \frac{(b-a)}{(b-c)}$$

5.1.10.4 Determination of Specific Gravity by Measurement of Displaced Water

In this method a graduated glass cylinder half-filled with water is taken. A previously weighted mineral piece is put inside the cylinder. The increase in volume of the water in the cylinder is noted. The specific gravity is computed by dividing the weight in grams of the mineral in air by increase in volume of the water in cm^3.

5.1.10.5 Determination of Specific Gravity by Pycnometer or Specific Gravity Bottle

In this case the weight of the mineral is first determined (W_1). The bottle is filled with distilled water. The combined weight of the mineral and the filled bottle is determined (W_2). The mineral is then put into the bottle from which it displaces an equal volume of water, and the weight is again determined (W_3). The weight of the water displaced is obtained by subtracting

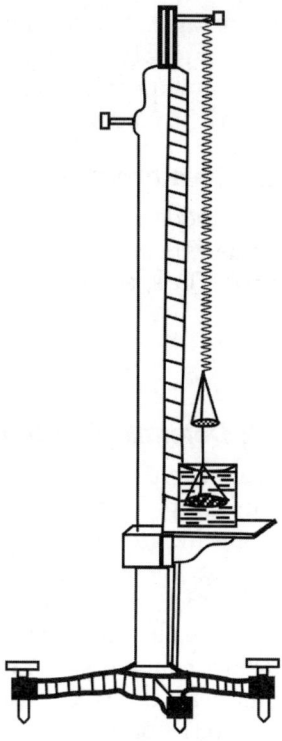

Fig. 5.2: Jolly's spring balance

the last weight from the preceding ($W_2 - W_3$). The specific gravity is obtained by dividing the weight of the mineral (W_1) by the weight of the water it displaces ($W_2 - W_3$).

5.1.10.6 Determination of Specific Gravity by Using Heavy Liquids

If two minerals of different specific gravities are placed in a liquid of intermediate specific gravity, the heavier mineral sinks to the bottom while the lighter mineral floats on the surface. By varying the specific gravity of the liquid it can be made equal to that of the mineral, when the mineral will remain in suspension, i.e. neither sink nor float. At this stage the mineral and the liquid have the same specific gravities. The common heavy liquids are bromoform and methylene iodide which have specific gravity 2.89 and 3.30 respectively. These two liquids mix well and can be diluted with acetone or benzene. The Clerici's solution is a saturated solution of equal parts of thallium formate and malonate in water. It has a specific gravity of 4.2 and it can be diluted with water. The heavy liquid is diluted until the mineral neither sinks nor floats in the liquid but remains in suspension. At this stage the specific gravity of the mineral is equal to that of the liquid. The specific gravity of the liquid is determined by pycnometer.

5.1.11 Taste, Odour and Feel

The water-soluble minerals have some characteristic tastes. The tastes given in Table 5.8 are noteworthy.

Table 5.8: Taste of certain minerals and substances

Taste	Substance	Taste	Substance
Saline	Halite, common salt	Alkaline	Potash and soda salts
Cooling	Potassium chlorate, Nitre	Astringent	Green vitriol
Sweet astringent	Alum	Bitter	Epsom salt, sylvite

Some minerals have characteristic odour that comes out when the mineral is rubbed, heated or blown with mouth. These are given in Table 5.9.

Table 5.9: Odour of certain minerals

Odour	Mineral
Earthy	Kaolin, bauxite when blown by mouth
Sulphurous	Sulphur, pyrites when struck, sulphides when heated
Garlic/alliaceous	Arsenic minerals when heated
Foetid (rotten egg smell)	Limestone and some quartz when heated or rubbed

Some minerals have characteristic feel as given in Table 5.10.

Table 5.10: Characteristic feel of certain minerals

Feel	Mineral	Feel	Mineral
Greasy	Graphite	Smooth	Galena
Soapy	Talc, chlorite	Rough	Bauxite

Some minerals like graphite and psilomelane mark the paper and soil fingers.

5.1.12 Tenacity

Some minerals have certain properties dependent upon their tenacity. These are given in Table 5.11.

Table 5.11: Tenacity property of certain minerals

Property	Description
Sectile	The mineral can be cut by a knife, e.g. graphite, gypsum, steatite
Malleable	The mineral flattens under hammer, e.g. native gold, silver and copper
Flexible	The mineral can be bent, e.g. talc, selenite
Elastic	The mineral restores its formal position after bent, e.g. mica flakes
Brittle	The mineral when struck yields powder in stead of slice

5.1.13 Magnetism, Electricity and Radioactivity

On the basis of magnetic strength, the minerals can be grouped under four groups given in Table 5.12.

Table 5.12: Magnetic property of certain minerals	
Degree of magnetism	*Mineral*
Highly magnetic	Magnetite, pyrrhotite
Moderately magnetic	Siderite, chromite, ilmenite, hematite
Weakly magnetic	Monazite, spinels, tourmaline
Non-magnetic	Quartz, feldspar, calcite and a number of other rock-forming minerals

Native metals like gold, silver and copper conduct electricity. Graphite, though a nonmetal, is a good conductor of electricity. Some minerals when heated up develop electric charge in different parts. Such minerals are known as *pyroelectric*. Tourmaline is a pyroelectric mineral. When heated up, the crystal becomes negatively charged at its sharp end and positively charged at its blunt end. When subjected to pressure, some minerals like quartz become electrically charged. This effect is known as *piezoelectricity*.

Minerals containing elements of high atomic weight are radioactive. They emit radiations of different degree. Radium, uranium and thorium are the chief radioactive elements.

5.1.14 Reaction with Acid

Some carbonate minerals like calcite, aragonite and dolomite (when finely powdered) react with hydrochloric acid. Such minerals are readily identified in the laboratory by their reactive property. Galena, sphalerite, etc. react with hydrochloric acid resulting in liberation of H_2S. Some of the minerals like sylvite, halite are soluble in nitric acid. A few other minerals are soluble in sulphuric acid or hydrofluoric acid.

5.2 MEGASCOPIC IDENTIFICATION OF MINERALS

Each mineral has its own set of physical properties by which it is identified in the laboratory and distinguished from others. A mineral may show some or all of the characters described above. For the convenience of description, the characters stated above have been categorized under two broad groups, viz. general and special/ specific properties. Physical properties like colour, streak, lustre, structure and form, cleavage, fracture, hardness and specific gravity are grouped under general category because they can be studied in case of all the minerals. The remaining properties like transparency, translucency, fluorescence, phosphorescence, taste, odour, feel, tenacity, reaction with acid and properties based on magnetism, electricity and radioactivity have been grouped under specific or special category because they are specific to some minerals. In the description table, the chemical composition, occurrence/association and use of each mineral have been given in addition to the general and special properties. All the properties of a mineral given in Table 5.13 are not necessarily diagnostic. The diagnostic properties have been indicated by '*' symbol.

Table 5.13: Megascopic properties of minerals

Mineral	Acanthite (Argentite)	Actinolite	Aegirine (Acmite)	Agate	Aikinite
Chemical composition	Ag_2S	$Ca_2(Fe, Mg)_5 Si_8O_{22}$ (OH)$_2$	Na (Al, Fe)Si_2O_6	SiO_2	$CuPbBiS_3$
Colour	Black*	White to green*	Brown or green*	White or irregularly clouded	Grey
Streak	Black*	White	Yellowish grey*	White	Grey-black*
Lustre	Metallic*	Vitreous	Vitreous	Vitreous	Metallic*
Structure and form	$4/m \bar{3} 2/m$ above 173°C, 2/m below 173°C filiform	2/m, prismatic* needle like* or columnar	2/m, slender prismatic*, fibrous aggregates	Cryptocrystalline*	2/m 2/m 2/m, acicular, massive
Cleavage	Imperfect	2 sets perfect at 56° and 124°	2 sets at 87° and 93° 1 set perfect	Absent*	Imperfect
Fracture	Conchoidal	Uneven	Uneven	Conchoidal	Subconchoidal
Hardness	2-2.5*	5-6	6-6.5	7*	2-2.5*
Specific gravity	7.1-7.3, very high*	3.0-3.3, moderate	3.4-3.55, moderate	2.6-2.64, low	6.1-6.7, very high
Special properties	Sectile* Marks paper*	Transparent to translucent	Translucent	Translucent	Opaque, reacts with HNO_3*
Occurrence	Veins associated with native silver	Greenschists, medium-grade metamorphic rocks	Soda rich and silica poor igneous rocks such as nepheline syenite	Filing cavities in amygdaloidal rocks	Hydrothermal quartz vein in association with sulphide minerals
Use	Ore of silver	Fibrous variety is used as asbestos		Semiprecious stone, manufacture of brooches, snuff-boxes and similar articles	Guide mineral for gold

Mineral	Allanite	Alunite	Amblygonite	Amphibole
Chemical composition	(Ca,Ce)$_2$ (Fe^{2+}, Fe^{3+}) Al_2O (SiO$_4$) (Si$_2O_7$) (OH)	KAl_3 (SO$_4$)$_2$ (OH)$_6$	$LiAlFPO_4$	Fe, Mg silicate
Colour	Brown to pitch-black*	White, grey, red	White to pale green or blue*, rarely yellow	Black*
Streak	Brownish black*	White, grey	White	White

(Contd...)

Lustre	Submetallic, pitchy, resinous*	Vitreous to pearly*	Vitreous, pearly on cleavage	Vitreous
Structure and form	2/m, slender to acicular prismatic or massive	3 m, usually massive, may be tabular	$\bar{1}$, cleavable masses*	Mostly crystalline, elongated*
Cleavage	Imperfect	2 sets, 1 set good	3 sets*	2 sets perfect at 55°*
Fracture	Subconchoidal	Conchoidal	Uneven	Uneven
Hardness	5.5-6	4*	6	5-6
Specific gravity	3.5-4.2, high	2.6-2.9, moderate	3-3.1, moderate	2.9-3.4, moderate
Special properties	Subtranslucent, brittle, gelatinizes in acids*	Transparent to translucent	Translucent	Translucent
Occurrence	Minor accessory of igneous rocks like granite, syenite, diorite, frequently associated with epidote	Associated with fumaroles or in zone of hydrothermal alteration in K-rich igneous rocks	Granite pegmatite	Igneous and metamorphic rocks
Use	Minor source of cesium	Production of alum, recovery of potassium and aluminium sulphate	Source of lithium	

Mineral	Analcite (Analcime)	Anatase (Octahedrite)	Andalusite	Anglesite
Chemical composition	$NaAlSi_2O_6 \cdot H_2O$	TiO_2	Al_2SiO_5	$PbSO_4$
Colour	Colourless, white, grey, pink*	Brown, blue or black*	Brown, red, green*	Colourless, white, grey
Streak	White	Colourless	White	White
Lustre	Vitreous	Adamantine*	Vitreous	Adamantine* when crystals, dull when massive
Structure and form	4/m $\bar{3}$ 2/m crystalline or massive	4/m 2/m 2/m generally crystallised	2/m 2/m 2/m elongate square prism*	2/m 2/m 2/m prismatic, massive, granular
Cleavage	1 set obscure	2 sets perfect	Imperfect	1 set perfect
Fracture	Uneven	Subconchoidal	Subconchoidal	Conchoidal
Hardness	5-5.5	5.5-6	7.5*	2.5-3*
Specific gravity	2.26-2.27, low	3.82-3.95, high*	3.16-3.2, moderate	6.2-6.4, high*
Special properties	Transparent to translucent, weakly pyroelectric*	Brittle, transparent to opaque	Transparent to translucent	Transparent to translucent
Occurrence	Alkaline igneous rock and also with zeolites	Hydrothermal veins	Regional metamorphic rocks	Oxidized Pb deposit

(Contd...)

Continued from previous page:

Use		Source of titanium	Source of mullite for manufacture of refractories and acid resistant articles also as a gemstone	Minor ore of lead

	Anhydrite	Ankerite	Anthophyllite	Apatite
Mineral	Anhydrite	Ankerite	Anthophyllite	Apatite
Chemical composition	$CaSO_4$	$CaFe(CO_3)_2$	$(Mg, Fe)_7Si_8O_{22}(OH)_2$	$Ca_5(PO_4)_3(OH, F, Cl)$
Colour	Colourless, bluish, violet	White, yellow, brown*	Brown, grey, green*	Green, yellow, blue
Streak	White	White	White	White
Lustre	Vitreous, pearly on cleavage*	Vitreous to pearly	Vitreous	Vitreous to subresinous*
Structure and form	2/m 2/m 2/m, prismatic, fibrous, granular, massive	$\bar{3}$, rhombohedral*, massive	2/m 2/m 2/m lamellar, bladed, fibrous*	6/m, prismatic*, granular, massive
Cleavage	3 sets perfect*	1 set perfect	2 set perfect* at 55°	2 sets, 1 set good
Fracture	Uneven	Subconchoidal	Uneven	Conchoidal
Hardness	3–3.5*	3.5*	5.5–6	5*
Specific gravity	2.89–2.98, moderate	2.95–3.1, moderate	2.9–3.2, moderate	3.2, moderate
Special properties	Transparent, soluble in HCl*	Transparent to translucent, powdered mineral reacts with HCl*	Transparent to translucent	Transparent to translucent
Occurrence	Associated with evaporite, gypsum, halite and cap rock of salt dome	Precambrian iron ore formations, replacement in limestones	Low grade Mg-rich metamorphic rocks	Accessory constituent of all types of rocks
Use	Soil conditioner, fertilizer, source of sulphur and production of plasters, cements, sulphates and H_2SO_4 also as ornamental stone		Amosite, a long fibrous variety of anthophyllite is used as asbestos	Fertilizer, gemstone and manufacture of phosphorus chemicals used in match industry

Mineral	Apophyllite	Aragonite	Arfvedsonite	Arsenopyrite
Chemical composition	$KCa_4Si_8O_{20}F \cdot 8H_2O$	$CaCO_3$	$Na_3Mg_4AlSi_8O_{22}(OH)_2$	$FeAsS$
Colour	Colourless, white, grey	Colourless, white, pale yellow	Black*	Silver white*
Streak	White	White	Bluish grey*	Black*
Lustre	Base—pearly, other faces—vitreous*	Vitreous	Vitreous	Metallic*

(Contd...)

Structure and form	4/m 2/m 2/m crystalline	2/m 2/m 2/m crystalline, tabular, radiating*, fibrous*	2/m, elongated prismatic*	2/m, prismatic, massive, granular
Cleavage	1 set perfect	2 sets, 1 set good	1 set perfect	Imperfect
Fracture	Uneven	Subconchoidal	Uneven	Uneven
Hardness	4.5–5*	3.5–4*	6	5.5–6*
Specific gravity	2.3–2.4, low	2.94, moderate	3.45, moderate	6.1, high*
Special properties	Transparent to translucent	Transparent to translucent, reacts with HCl*	Translucent	Opaque
Occurrence	Associated with zeolites, lining cavities in basalt	Disseminated in gypsum beds, sedimentary iron ores, hot spring deposits, pearls	Na rich igneous rocks like nepheline-syenite	Veins, pegmatites, contact aureoles, disseminations in low to moderate metamorphic rocks
Use				Source of arsenci compounds used in the manufacture of dyes, chemical and leather industries

Mineral	Atacamite	Augite	Autunite	Axinite
Chemical composition	$Cu_2Cl(OH)_3$	(Ca, Mg, Fe, Na) (Mg, Al, Fe) (Si, Al)$_2$O$_6$	$Ca(UO_2)_2(PO_4)_2 \cdot 10H_2O$	(Ca,Fe^{2+}, Mn)$_3$ Al$_2$ (BO)$_3$ (Si$_4$O$_{12}$) (OH)
Colour	Shades of green*	Black*, greenish black	Lemon yellow to pale green*	Clove brown, grey, violet, green, yellow*
Streak	Green*	White	Yellow*	Colourless
Lustre	Adamantine* to vitreous	Vitreous	Pearly on basal pinacoid and vitreous on other faces*	Vitreous
Structure and form	2/m 2/m 2/m, prismatic, fibrous, granular, massive	2/m, crystalline, prismatic*, massive	4/m 2/m 2/m, tabular, foliated, scaly	$\bar{1}$, thin and sharp edged crystal*, lamellar, massive, granular
Cleavage	1 set perfect	2 sets perfect at 90°*	1 set perfect	1 set perfect
Fracture	Conchoidal	Uneven	Uneven	Conchoidal
Hardness	3–3.5*	5–6*	2–2.5*	6–7
Specific gravity	3.76, high	3.2–3.4, moderate	3.15, moderate	3.27–3.35, moderate

(Contd...)

Special properties	Transparent to translucent	Transparent to translucent	Transparent to translucent, strongly fluorescent in ultraviolet light*	Transparent to translucent
Occurrence	Supergene mineral in oxidized zones of copper deposits	Mafic and intermediate igneous rocks	Secondary alteration zone of uraninite	Cavities in granite and contact zone surrounding granitic intrusions, contact metamorphic zones
Use	Minor ore of copper		Ore of uranium	

Mineral	Azurite	Barite (Barytes)	Bauxite
Chemical composition	$Cu_3(CO_3)_2(OH)_2$	$BaSO_4$	Mixture of diaspore ($HAlO_2$), bohemite [$AlO(OH)$], gibbsite [$Al(OH)_3$]
Colour	Azure blue*	Colourless, white, grey	Dirty white, greyish, reddish brown
Streak	Blue*	White	Greyish white*
Lustre	Vitreous, dull, earthy	Vitreous, pearly on basal pinacoid*	Dull and earthy*
Structure and form	2/m, massive, rarely tabular or prismatic	2/m 2/m 2/m, tabular, massive, concretions	Amorphous, pisolitic*, massive
Cleavage	2 sets, 1 set perfect	3 sets, 2 set perfect*	Absent*
Fracture	Conchoidal	Uneven	Uneven
Hardness	3.5–4*	3–3.5*	1–3*
Specific gravity	3.77, high*	4.5, high*	2–2.55, low
Special properties	Transparent to translucent, brittle, reacts with HCl*	Transparent to translucent	Translucent, soils hand, earthy smell when moistened with water*
Occurrence	Secondary copper mineral formed by alteration of copper oxides and sulphides	Hydrothermal vein, residual mass in clay overlying limestone, hot springs	Chemical weathering of aluminous rocks, residual deposit, transported
Use	Minor copper ore and manufacture of blue pigment	Manufacture of white paint, dressing of poor-quality calico, etc., heavy mud in drilling, source of barium; fireworks, sugar refining, making enamels, preparation of high refractive index glass, medicine, paint, rubber and paper industries	Manufacture of Al_2O_3 from which Al metal is extracted and other Al-compounds are prepared, abrasive, refractory, etc.

(Contd...)

Mineral	Beryl	Biotite	Bismuthinite	Boracite
Chemical composition	$Be_3Al_2Si_6O_{18}$	$K(Mg, Fe)_3(AlSi_3O_{10})(OH)_2$	Bi_2S_3	$Mg_3ClB_7O_{13}$
Colour	Bluish green* to light yellow, pink, white	Brown, black*	Lead-grey*	White, grey, yellow, greenish*
Streak	White	White	Lead-grey*	White
Lustre	Vitreous	Splendent, pearly*	Metallic	Vitreous
Structure and form	6/m 2/m 2/m, hexagonal prismatic* columnar	2/m, sheet like*, foliated*, book like*	2/m 2/m 2/m, Massive, foliated or fibrous	4/m $\bar{3}$ 2/m, crystalline*, columnar, massive
Cleavage	Imperfect	1 set perfect*	3 sets, 1 set perfect	Imperfect
Fracture	Even	Ragged	Uneven	Conchoidal, uneven
Hardness	7.5–8*	2.5–3*	2*	7*
Specific gravity	2.7–2.9, moderate	2.8–3.2, moderate	6.4–6.5, very high	2.95, moderate
Special properties	Transparent to translucent	Transparent, elastic*	Sectile*, Opaque	Subtransparent to subtranslucent
Occurrence	Granite pegmatite	Igneous, pegmatite, metamorphic; (vermiculite is formed by alteration of biotite)	Hydrothermal vein and contact metasomatic deposits	Saline deposit in association with rock-salt, gypsum and anhydrite
Use	Major source of Be, transparent varieties are gemstones: Emerald—deep green, aquamarine—greenish blue, morganite—pink	Insulator in electrical industry, fire proof material	Chief source of bismuth, production of fusible alloys, glass with high birefringence, chemicals, medical preparations	Source of boron

Mineral	Borax	Bornite	Boulangerite	Bournonite
Chemical composition	$Na_2B_4O_5(OH)_4 \cdot 8H_2O$	Cu_5FeS_4	$Pb_5Sb_4S_{11}$	$CuPbSbS_3$
Colour	Colourless, white, grey	Brownish bronze*	Lead-grey to black*	Steel-grey to black*
Streak	White	Greyish black*	Greyish-black*	Steel-grey to black*
Lustre	Vitreous, resinous	Metallic*	Metallic*	Metallic*
Structure and form	2/m, prismatic, massive, cellular	$\bar{4}$2m, crystalline, massive granular aggregates	2/m, crystalline, massive granular aggregate	2/m 2/m 2/m, prismatic, massive, granular
Cleavage	1 set perfect	Imperfect	1 set distinct	Imperfect
Fracture	Conchoidal	Conchoidal	Conchoidal, uneven	Conchoidal, uneven

(Contd...)

Hardness	2–2.5*	3*	2.5–3*	2–3*
Specific gravity	1.7–1.9, low*	5, high*	6.23, very high*	5.7–5.9, high*
Special properties	Translucent, brittle, soluble in water*, sweetish-alkaline taste*	Opaque	Opaque	Opaque, brittle
Occurrence	Associated with evaporite deposit	Sulphide veins, secondary enrichment zone	Hydrothermal veins	Hydrothermal veins
Use	Antiseptic and preservative; soldering and welding; flux, deoxidiser, neutron absorber, additive in fuel; boron carbide as abrasive.	Ore of copper	Ore of lead	Ore of lead and copper

Mineral	Braunite	Brookite	Brucite
Chemical composition	Mn_2O_3	TiO_2	$Mg(OH)_2$
Colour	Black	Brown, reddish*	White, green, grey
Streak	Brownish black	Colourless to greyish	White
Lustre	Submetallic	Submetallic*	Pearly on base, other sides vitreous to waxy*
Structure and form	4/m, dipyramidal and massive	2/m 2/m 2/m, crystalline, prismatic*	$\bar{3}$ 2/m, crystalline, foliated*, massive
Cleavage	1 set distinct	Imperfect	1 set perfect (basal)
Fracture	Uneven	Conchoidal, uneven	Conchoidal, uneven
Hardness	6–6.5	5–6	2.5*
Specific gravity	4.7–5, high	3.87–4.14, high*	2.4–2.5
Special properties	Opaque	Translucent, opaque, brittle	Transparent, translucent, sectile*
Occurrence	Contact metasomatic and hydrothermal veins	Hydrothermal veins, alteration product of other Ti-bearing minerals	Alteration product of periclase, serpentine; also in crystalline limestone
Use	Important ore of manganese; used for making of ferromanganese; flux in steel smelting	Source of titanium	Refractory, source of magnesium

(Contd...)

Mineral	Calcite	Cancrinite
Chemical composition	$CaCO_3$	$[Na_3Ca(AlSiO_4)_3(CO_3, SO_4)]$
Colour	Colourless to grey, differently coloured when impure	Yellow*, white, red, blue, etc.
Streak	White	Colourless
Lustre	Vitreous, pearly, earthy*	Vitreous on cleavage surface and greasy on fracture
Structure and form	$\bar{3}$ 2/m, crystalline, rhombohedral*, granular, stalactitic	6, prismatic, massive
Cleavage	3 sets perfect* rhombohedral	1 set perfect
Fracture	Conchoidal	Conchoidal, uneven
Hardness	3* on cleavage and 2.5 on base	5–5.5*
Specific gravity	2.71	2.42–2.48
Special properties	Transparent (iceland spar), translucent, reacts with dil. HCl*	Translucent, brittle
Occurrence	Limestone, cave deposit, carbonatite, calcareous marl and sandstone	In association with nepheline seynite and other undersaturated rocks
Use	Manufacture of cement, quicklime, bleaching powder, calcium carbide, glass, soap, paper, paints, chalk, fertilizer; iceland spar is used for manufacture of Nicol prism; used as flux in blast furnace, building stone for statuary, switchboard in electrical industry; satin spar and onyx are used as gem materials	

Mineral	Cassiterite	Celestite	Celsian	Cerargyrite	Cerussite
Chemical composition	SnO_2	$SrSO_4$	$BaAl_2Si_2O_8$	$AgCl$	$PbCO_3$
Colour	Brown, black*	Colourless, white, faint blue, light red	Colourless*	Colourless, pale shades of grey	Colourless, grey, white
Streak	White	White	White*	Shining white*	White
Lustre	Adamantine* to submetallic	Vitreous, pearly*	Vitreous	Resinous, adamantine*	Adamantine*
Structure and form	4/m 2/m 2/m, pyramidal*, massive, colloform, reniform, fibrous	2/m 2/m 2/m, tabular, acicular*, reniform*, granular	2/m, prismatic	4/m $\bar{3}$ 2/m, massive, wax like*, columnar	2/m 2/m 2/m, tabular, prismatic, acicular, granular, massive, platy crystals*
Cleavage	Imperfect	3 sets*, 1 set good	2 sets perfect	Imperfect	2 sets, 1 set good
Fracture	Subconchoidal	Uneven	Conchoidal	Conchoidal	Conchoidal

(Contd...)

Hardness	6–7	3–3.5	6–6.5	2–3*	3–3.5*
Specific gravity	7, high*	3.97, high*	3.37	5.5, high*	6.55, high*
Special properties	Translucent	Transparent to translucent	Transparent to translucent	Transparent to translucent, sectile*	Transparent to translucent, reacts with HNO_3*
Occurrence	Pegmatites, contact aurioles, placers	Disseminated in limestone and sandstone	Contact metasomatic zone	Gossans, upper parts of silver vein	Supergene lead ore
Use	Important ore of tin; manufacture of ceramics, enamel and pigment	Chief source of strontium salts for fireworks, sugar refining, manufacture of iridescent glass, specially glazed bricks	Source of barium, manufacture of glass and ceramics	Ore of silver	Important ore of lead; production of white lead, which is used as pigment

Mineral	Chabazite	Chalcanthite	Chalcedony	Chalcocite
Chemical composition	$CaAl_2Si_4O_{12} \cdot 6H_2O$	$CuSO_4 \cdot 5H_2O$	SiO_2	Cu_2S
Colour	Colourless, white, red, yellow	Sky blue* to greenish	White, grey	Lead-grey*
Streak	White	Uncoloured	White	Greyish black*
Lustre	Vitreous	Vitreous	Vitreous, wax like*	Metallic*
Structure and form	$\bar{3}$ 2/m, crystalline, rhombohedral*	1, massive, stalactitic, reniform	Cryptocrystalline*, mammillary, botryoidal, stalactitic*	2/m 2/m 2/m, tabular, massive
Cleavage	Imperfect	Imperfect	Absent*	Imperfect
Fracture	Uneven	Conchoidal	Conchoidal	Conchoidal
Hardness	4–5*	2.5*	7*	2.5–3*
Specific gravity	2.1, low	2.1–2.3	2.6, low	5.5–5.8, high*
Special properties	Transparent to translucent, brittle, gelatinise in HCl*	Opaque, brittle	Transparent to translucent	Opaque, imperfectly sectile*
Occurrence	Secondary mineral formed in cracks and cavities with other zeolite minerals	Oxidized zone of copper sulphide deposits	Lining or filling cavities in rocks	Supergene enriched zone
Use		Used as pesticide and dye	Different varieties of chalcedony are used as gemstones	Copper ore

(Contd…)

Mineral	Chalcopyrite	Chloanthite	Chlorite	Chloritoid
Chemical composition	$CuFeS_2$	$NiAs_{3-x}$	$(Mg, Fe, Al)_6 (Si, Al)_4O_{10} (OH)_8$	$(Fe, Mg) Al_2SiO_5 (OH)_2$
Colour	Brass-yellow*	Tin-white, grey*	Green* of various shades	Dark green*
Streak	Greenish black*	Greyish-black*	Greenish white*	Colourless
Lustre	Metallic*	Metallic	Vitreous to pearly*	Pearly*
Structure and form	$\bar{4}2m$, crystalline, massive aggregates	$2/m\,\bar{3}$, cubic, massive granular	$2/m$, micaceous habit*, foliated books	$2/m$, scaly, platy, foliated*
Cleavage	Imperfect	2 sets distinct	1 set perfect	1 set good
Fracture	Uneven	Conchoidal	Subconchoidal	Subconchoidal
Hardness	3.5–4*	5.5–6	2–2.5*	6.5*
Specific gravity	4.1–4.3, high*	6.4–6.7, high	2.6–3.3, moderate	3.5–3.8, high
Special properties	Opaque, brittle	Subtranslucent to opaque	Transparent to translucent, greasy feel*, gelatinize in HCl*, decomposed by boiling with conc. H_2SO_4*, scratched by finger nail*	Transparent to translucent
Occurrence	Hydrothermal vein, replacement deposit	Hydrothermal deposits	Low to intermediate grade metamorphic rocks (greenschist facies)	Low to intermediate grade metamorphic rocks rich in Al and Fe
Use	Chief copper ore	Minor ore of nickel and arsenic	Building stone	

Mineral	Chondrodite	Chromite	Chrysoberyl (Alexandrite)
Chemical composition	$Mg_5 (SiO_4)_2(OH, F)_2$	$FeCr_2O_4$	$BeAl_2O_4$
Colour	White, yellow, red*	Brownish black*	Yellow, green, brown*
Streak	White	Brown*	White*
Lustre	Vitreous to resinous*	Metallic to submetallic	Vitreous
Structure and form	$2/m$, crystalline, massive	$4/m\,\bar{3}\,2/m$, octahedral*, granular, massive	$2/m\,2/m\,2/m$, crystalline*, tabular, prismatic
Cleavage	Imperfect	Absent	1 set good
Fracture	Subconchoidal	Conchoidal, uneven	Subconchoidal, uneven
Hardness	6–6.5*	5.5	8.5*

(Contd...)

Specific gravity	3.16–3.26, moderate	4.6–5.1, high*	3.65–3.8, high
Special properties	Transparent to translucent	Subtranslucent to opaque, slightly magnetic*	Transparent to translucent
Occurrence	Metamorphosed carbonate rocks, rarely in carbonatites	Peridotite and other ultrabasic rocks	Rare mineral in granites, pegmatites, placers
Use		Chief ore of chromium, raw material for making stable paints, tanning agents and chemicals like potassium dichromates manufacture of refractory bricks, etc.	Coloured and transparent varieties are used as gemstone. Alexandrite—emerald-green (red in artificial light) and cymophane (cat's eye)—green chatoyant gemstone

Mineral	Chrysocolla	Chrysotile	Cinnabar
Chemical composition	$Cu_4H_4Si_4O_{10}(OH)_8$	$Mg_6Si_4O_{10}(OH)_8$	HgS
Colour	Green to greenish blue*	White to greenish white*	Bright red to brownish red*
Streak	White	White	Scarlet*
Lustre	Vitreous to earthy	Greasy to waxy in massive and silky in fibrous variety*	Adamantine*
Structure and form	Cryptocrystalline* to amorphous	2/m, fibrous, asbestiform*	32, fine granular, massive, earthy, acicular
Cleavage	Absent	Absent	1 set perfect
Fracture	Conchoidal	Uneven	Subconchoidal
Hardness	2–4*	3–5*	2.5*
Specific gravity	2–2.4, low	2.5–2.6, moderate	8.1, very high*
Special properties	Translucent to opaque, sectile*	Translucent	Transparent to translucent, marks paper*
Occurrence	Oxidized zone of copper deposits	Secondary mineral in mafic and ultramafic igneous rock	Vein fillings near volcanic rock and hot springs
Use	Minor ore of copper	Chief source of asbestos, massive translucent green variety is used as ornamental stone and building material; manufacture of fire-proof suits and curtains. As filters, brake bands and asbestos-rubber articles; refractory, light-weight roofing, pipes, cardboard, heat- and electricity-insulating paper, gaskets; heat-resistant paints, boiler insulation, plaster and other materials.	Important source of mercury; separation of gold, manufacture of chemicals, high explosives, scientific instruments; preparation of paint, vermilion, etc.

(Contd...)

Mineral	Clinozoicite	Cobaltite	Colemanite
Chemical composition	$Ca_2Al_3Si_3O_{12}(OH)$	$(Co, Fe)AsS$	$CaB_3O_4(OH)_3 \cdot H_2O$
Colour	Light green, yellow, grey*	Tin-white to silver-white*	Colourless* to white, grey
Streak	White	Greyish black*	White
Lustre	Vitreous	Metallic*	Vitreous
Structure and form	2/m, prismatic*, fibrous, acicular, granular, massive	mm 2, massive, granular	2/m, prismatic, massive, granular
Cleavage	1 set perfect	1 set good	1 set perfect
Fracture	Uneven	Uneven	Subconchoidal
Hardness	6–7*	5.5	4–4.5*
Specific gravity	3.1–3.4, moderate	6.3, high*	2.42, low
Special properties	Transparent to translucent	Opaque	Transparent to translucent
Occurrence	Metamorphosed Ca rich and Fe poor rocks	Disseminations in high temperature metamorphic rocks	Inter-stratified with lake bed deposits of Tertiary age
Use	The rose-red variety (thulite) is used as gemstone.	Chief source of cobalt; preparation of deep-blue and green pigments used in glass staining; tinting of ceramics; manufacture of superior grade steel	A source of borax

Mineral	Columbite - Tantalite	Copper	Cordierite
Chemical composition	$(Fe, Mn)(Nb, Ta)_2O_6$	Cu	$(Mg, Fe)_2Al_4Si_5O_{18} \cdot nH_2O$
Colour	Iron black to brown*	Copper-red*	Indigo to greyish blue*
Streak	Brown to black*	Copper-red*	White
Lustre	Submetallic*	Metallic*	Vitreous
Structure and form	2/m 2/m 2/m, crystalline	$4/m \bar{3} 2/m$, crystalline, plates, scales	2/m 2/m 2/m, crystalline, granular*, massive
Cleavage	1 set good	Absent	2 sets, 1 set good
Fracture	Subconchoidal	Hackly*	Subconchoidal
Hardness	6	2.5–3*	7–7.5*
Specific gravity	5.2–7.9, high	8.7–8.9, very high*	2.5–2.8, moderate
Special properties	Translucent to opaque	Opaque, highly ductile* and malleable*	Transparent to translucent
Occurrence	Granitic rocks and pegmatite	Oxidized zone of copper deposits, hydrothermal deposit	High grade metamorphosed aluminous rock

(Contd...)

Mineral	Corundum	Covellite	Cristobalite	Crocoite
Use	Source of tantalum and niobium for production of special steel	Minor ore of copper, alloys: Brass (Cu + Zn), bronze (Cu + Sn), German silver (Cu + Zn + Ni)		Transparent variety (dichroite) is used as gemstone
Chemical composition	Al_2O_3	CuS	SiO_2	$PbCrO_4$
Colour	Colourless, blue, red, orange, yellow	Indigo-blue*	Colourless*	Red*
Streak	White	Dark grey to black*	White	Orange-yellow*
Lustre	Adamantine* to vitreous	Metallic*	Vitreous	Adamantine
Structure and form	$\bar{3}$ 2/m, crystalline, prismatic*, granular, massive	6/m 2/m 2/m, massive, foliated*	Low cristobalite—422 High cristobalite—4/m $\bar{3}$ 2/m	2/m, slender prismatic, striated, granular
Cleavage	Absent	1 set perfect	Absent	Imperfect
Fracture	Uneven	Conchoidal	Conchoidal*	Conchoidal to uneven
Hardness	9*	1.5–2*	6.5–7	2.5–3*
Specific gravity	3.9–4.1, high*	4.6–4.7, high*	2.32–2.33, low	6, high*
Special properties	Transparent to translucent	Opaque	Transparent to translucent	Translucent
Occurrence	Silica deficient igneous rocks, crystalline limestone, mica-schist, gneiss, etc.	Supergene mineral in copper deposits	High temperature silicic volcanic rocks	Oxidised zones of Pb deposit
Use	Sapphire (blue) and ruby (red) are gem varieties of corundum; Emery, greyish-black variety of corundum and alundum, made by fusing corundum with bauxite are used as abrasive	Minor ore of copper	Same as quartz	Minor ore of lead; chromium was first discovered from crocoite

Mineral	Cryolite	Cubanite	Cummingtonite	Cuprite
Chemical composition	Na_3AlF_6	$CuFe_2S_3$	$(Mg, Fe)_7Si_8O_{22}(OH)_2$	Cu_2O
Colour	Colourless to snow white*	Bronze yellow*	White, green*	Various shades of red*
Streak	White	Brownish*	White	Brownish red*
Lustre	Pearly*, greasy*, vitreous	Metallic	Vitreous; silky in fibrous variety*	Metallic to adamantine*

(Contd...)

Structure and form	2/m, massive, lamellar, columnar*	2/m 2/m 2/m, tabular, massive	2/m, prismatic, fibrous, lamellar*	4/m $\bar{3}$ 2/m, crystalline, crystallised
Cleavage	Absent	Absent	2 sets perfect	Imperfect
Fracture	Uneven	Uneven	Uneven	Conchoidal
Hardness	2.5*	3.5*	5.5–6*	3.5–4*
Specific gravity	2.95–3, moderate	4.7, high	2.9–3.2, moderate	5.9–6.1, high*
Special properties	Transparent to translucent, invisible when immersed in water*	Highly magnetic*	Translucent	Opaque
Occurrence	Granitic rocks	In association with chalcopyrite	Medium grade metamorphic rocks, amphibolites	Secondary mineral in the oxidized zone of copper deposit
Use	Manufacture of milk-white glass, porcelain, as a flux for cleansing metal surfaces, enamel for kitchenware and extraction of aluminium	Minor ore of copper	Amosite, an ash-grey coloured, asbestiform variety is used as asbestos	Minor ore of copper

Mineral	Datolite	Diamond	Diopside	Dolomite	Enargite
Chemical composition	CaB(SiO$_4$) (OH)	C	CaMgSi$_2$O$_6$	CaMg (CO$_3$)$_2$	Cu$_3$AsS$_4$
Colour	White, greyish, greenish, yellow, red*	Colourless, pale yellow, red, orange, green, blue, brown	White to light green*	White, colourless, shades of pink	Greyish black, iron-black*
Streak	White	White	White, light green	White	Greyish black*
Lustre	Vitreous	Adamantine*, uncut-greasy*	Vitreous	Vitreous to glistening*, pearly in some cases	Metallic*
Structure and form	2/m, crystalline, massive	4/m $\bar{3}$ 2/m, crystallized*	2/m, crystalline, prismatic*, massive	$\bar{3}$, rhombohedral*, massive, granular	mm2, crystalline, massive, columnar
Cleavage	Absent	4 sets perfect	2 sets perfect at 90°*	3 sets perfect*, rhombohedral	2 sets perfect
Fracture	Conchoidal to uneven	Conchoidal	Uneven	Subconchiodal	Uneven
Hardness	5–5.5*	10*	5–6	3.5–4	3*
Specific gravity	2.8–3, moderate	3.5, moderate	3.2–3.6, moderate	2.85, moderate	4.45, high*

(Contd...)

Special properties	Transparent to translucent	Transparent, high refractive index* (2.42)	Transparent to translucent	Transparent to translucent, reacts with hot HCl*	Opaque
Occurrence	Secondary mineral found in cavities of basalt lava	Altered ultramafic rocks (kimberlite), alluvial placers	Mafic and ultramafic igneous rocks; medium–high grade metamorphic rocks	Carbonate sediments, marble, limestone, hydrothermal veins	Vein and replacement sulphide deposits
Use	Source of boron	Gem, glass cutting, abrasive, drilling bits	Transparent varieties as gemstone	Building and ornamental stone, manufacture of cement, refractory, flux, ore of metallic magnesium and source of CO_2	An ore of copper, arsenic oxide is also extracted

Mineral	Enstatite	Epidote	Epsomite	Erythrite
Chemical composition	$MgSiO_3$	$Ca_2(Al, Fe)_3Si_3O_{12}(OH)$	$MgSO_4 \cdot 7H_2O$	$Co_3(AsO_4)_2 \cdot 8H_2O$
Colour	Greyish, greenish-white, brown*	Pistachio-green*, yellowish green, black	Colourless, white*	Crimson, pink, purple-red*
Streak	White, grey	White	White	Pale purple*
Lustre	Vitreous to pearly*	Vitreous	Vitreous	Adamantine* to vitreous, pearly on cleavage
Structure and form	2/m 2/m 2/m, prismatic, massive, blocky, fibrous	2/m, prismatic, fibrous, acicular, elongated	222, prismatic, fibrous, botryodal*, colloform*	2/m, prismatic, acicular, reniform, globular
Cleavage	2 sets perfect at 90°*	2 sets, 1 set perfect*	1 set perfect	1 set perfect
Fracture	Uneven	Uneven	Conchoidal	Uneven
Hardness	5.5–6	6–7	2–2.5*	1.5–2.5*
Specific gravity	3.2–3.6, moderate	3.4–3.5, high	1.68, low*	3.06, moderate
Special properties	Translucent	Transparent to translucent	Transparent to translucent. Taste-bitter and saline*, scratched by finger nail*, soluble in water and acids*	Transparent to translucent, sectile*
Occurrence	Mafic igneous rocks, high-grade metamorphic rocks	Low-to medium-grade metabasites and marbles	Caves, mine adits as encrustation, precipitates on carbonate or mafic igneous rocks	As cobalt bloom on other cobalt minerals

(Contd...)

Use	Gemstone	Tanzanite, a blue epidote is used as gemstone	Used in textile, paper making, sugar refining, chemical, pharmaceutical industries and tanning	Path finder for cobalt minerals and native silver

Mineral		Fluorite	Franklinite	Galena
Chemical composition		CaF_2	$(Zn, Fe, Mn)(Fe, Mn)_2O_4$	PbS
Colour		Colourless to light green*, yellow, purple	Black, reddish brown*	Lead-grey*
Streak		White	Redish-brown*	Lead-grey*
Lustre		Vitreous	Metallic	Bright metallic*
Structure and form		$4/m\,\bar{3}\,2/m$, crystalline-cubic*, massive, granular, columnar	$4/m\,\bar{3}\,2/m$, crystalline, rounded grains, massive	$4/m\,\bar{3}\,2/m$, crystalline—cubic*, massive, granular
Cleavage		4 sets* perfect (octahedral)	Absent	3 sets perfect*
Fracture		Conchoidal	Conchoidal	Subconchoidal, even
Hardness		4*	6	2.5*
Specific gravity		3.18, moderate	5.32, high*	7.6, very high*
Special properties		Transparent to translucent, fluorescent*	Opaque, slightly magnetic*	Opaque, reacts with HCl*
Occurrence		Hydrothermal vein, replacement deposit	Zn ore deposits	Hydrothermal veins, sulphide deposits
Use		Flux in steel making, foundry work; enameling iron, manufacture of opaque and opalescent glass, preparation of hydrofluoric acid, transparent variety is used for manufacture of microscope lenses and prisms	Minor ore of zinc	Ore of lead, preparation of pigment and source of silver

Mineral	Garnet	Glaucophane	Goethite
Chemical composition	$(Ca, Mg, Fe/Mn)_3(Fe, Al, Cr)_2(SiO_4)_3$	$Na_2Mg_3Al_2Si_8O_{22}(OH)_2$	$FeO(OH)$
Colour	Brown*, red*, yellow, white, green, black	Blue, grey*	Yellowish brown to dark brown*
Streak	White	White to light blue*	Brownish yellow*
Lustre	Vitreous to resinous*	Vitreous	Subadamantine to earthy
Structure and form	$4/m\,\bar{3}\,2/m$, crystallized*, dodecahedral*, massive granular, rounded grains	$2/m$, acicular, fibrous*, asbestiform*	$2/m\,2/m\,2/m$, prismatic, fibrous, bottryoidal or mammilary*, concentric growth bands common*

(Contd...)

Cleavage	Absent*	2 sets perfect at 60°*	Imperfect
Fracture	Subconchoidal to uneven	Uneven	Uneven
Hardness	6.5–7.5*	6–6.5	5–5.5*
Specific gravity	3.5–4.3, high*	3.1–3.2, moderate	4.3, high*
Special properties	Transparent to translucent	Transparent to translucent	Subtranslucent to opaque, magnetic when heated
Occurrence	Granite pegmatite, peridotite, kimberlite, metamorphic rocks	Blueschist facies metamorphic rock, marble	Weathering product of iron-bearing minerals; it forms the gossan (iron hat) over metalliferous veins
Use	Manufacture of garnet paper for polishing of hard wood, glass plate, for finishing leather, hard rubber and other articles; rhodolite (rose red pyrope), almandite, demantoid (green andradite) and essonite are used as gemstones		A minor source of iron

Mineral	Gold	Graphite	Greenockite	Grunerite
Chemical composition	Au	C	CdS	$Fe_7Si_8O_{22}(OH)_2$
Colour	Golden yellow*	Lead-grey, black*	Honey-coloured,* Orange-yellow	Dark green, brown*
Streak	Golden yellow*	Black and shining*	Orange-yellow to brick red*	White*
Lustre	Metallic*	Submetallic*	Adamantine, resinous	Vitreous, silky*
Structure and form	$4/m\,\bar{3}\,2/m$, filiform*, reticulated, nuggets*, grains, wire, scales,	$\bar{3}\,2/m$, foliated*, scaly, radiated, granular, books	6, barrel-shaped, acute pyramidal	$2/m$, fibrous, columnar, bladed, radiating*
Cleavage	Absent	1 set perfect (basal)	1 set perfect	2 sets perfect
Fracture	Hackly*	Subconchoidal	Subconchoidal	Uneven
Hardness	2.5–3*	1–2*	3–3.5*	6*
Specific gravity	15.6–19.3, very high*	2.1–2.2, low	5, high	3.1–3.6, moderate
Special properties	Opaque, malleable*, ductile*	Sectile*, marks paper*, soils hand*, good conductor of heat and electricity*	Transparent to translucent, brittle	Transparent to translucent
Occurrence	Veins related to silicic igneous rocks, hydrothermal gold-quartz vein, placer deposits	Metamorphic rocks like schists, marbles and gneisses	In association with sphalerite, wurtzite, etc.	Regionally metamorphosed Fe-rich sediment

(Contd...)

Property				
Use	Monetary standard, jewellery, scientific instruments, electroplating, gold leaf, dental appliances	Refractory crucibles, lead of pencils, paint, battery, electrodes, generator brush, lubricant, stovepolish, pigments, etc.	Used for electroplating, anticorrosion coating on iron and steel; preparation of storage battery, manufacture of automatic fire-fighting equipment.	Amocite, an asbestiform variety with ash-grey colour and long flexible fibers is used as asbestos
Mineral	Gypsum	Halite	Hausmannite	
Chemical composition	$CaSO_4 \cdot 2H_2O$	$NaCl$	Mn_3O_4	
Colour	Colourless, white, grey, brown, red	Colourless, white, differently coloured when impure	Black*	
Streak	White	White	Brown*	
Lustre	Vitreous, pearly, shining, silky*	Vitreous	Adamantine to submetallic	
Structure and form	$2/m$, tabular and platy crystals, acicular, compact*, massive*, granular, foliated	$4/m\,\bar{3}\,2/m$, cubic*, massive, granular	$4/m\,2/m\,2/m$, massive, granular	
Cleavage	2 sets, basal set perfect	3 sets perfect*	1 set perfect	
Fracture	Conchoidal	Conchoidal, uneven	Uneven	
Hardness	2*	2.5*	5–5.5	
Specific gravity	2.32, low	2.16, low	4.8, high	
Special properties	Transparent to translucent, greasy feel*, scratched by finger nail*, soluble in hot dil. HCl*	Transparent to translucent, salty taste*, diathermanous, soluble in water* and HNO_3*	Opaque	
Occurrence	Evaporite deposits, occasionally in fumaroles	Evaporite deposits, salt flats, salt domes, sedimentary beds, deposits from volcanic gases	Contact metasomatic and hydrothermal veins	
Use	Production of plaster of Paris, soil conditioner, fertilizer, cement; satin spar, filler for heavy white paper, cryons, paint, rubber etc; adulterant of foods; flux in smelting oxidized nickel ores. Alabaster, a variety of gypsum is used for ornamentation purpose.	Used in diet and food-preserving; absorber of moisture and oxygen in the purification of noble gases; source of Na and Cl, manufacture of HCl and preparation of sodium compounds particularly $Na_2(CO_3)$ for glass-making, soap-making, etc.	Manganese ore and used for making ferromanganese	
Mineral	Hauyne	Hedenbergite	Hematite	
Chemical composition	$Na_6Ca\,(AlSiO_4)_6\,(SO_4)$	$CaFeSi_2O_6$	Fe_2O_3	

(Contd...)

Property			
Colour	Blue, yellow, red*	Black*	Steel-grey, iron-black*; (red earthy variety—red ocher)
Streak	Colourless*	White*	Cherry red*
Lustre	Vitreous, greasy on fracture	Vitreous	Metallic*
Structure and form	$4/m\,\bar{3}\,2/m$, octahedra*, massive, granular	$2/m$, crystalline, massive	$\bar{3}\,2/m$, rhombohedral, massive, rosettes, botryodal, reniform, micaceous, foliated; (Platy metallic variety—Specularite)
Cleavage	1 set distinct	2 sets perfect at 87° and 93°	Absent
Fracture	Subconchoidal to uneven	Uneven	Subconchoidal to uneven
Hardness	5.5–6	5–6*	5.5–6.5
Specific gravity	2.4–2.5, low	3.55, high	4.9–5.3, high*
Special properties	Translucent	Transparent to translucent	Translucent to opaque, magnetic on heating*
Occurrence	Silica poor volcanic rocks in association with nepheline and leucite	Skarn mineral, occurs at the contact zone of granitic rocks and limestone; mafic and ultramafic igneous rocks; medium—high grade metamorphic rocks	Sedimentary iron ore, contact metamorphic deposit, accessory mineral of igneous rock, regionally metamorphosed rock, red sandstone as cement
Use			Ore of iron, used as pigment, manufacture of red pencil and polishing powder

Mineral	Hemimorphite (Calamine)	Heulandite	Hornblende
Chemical composition	$Zn_4(Si_2O_7)(OH)_2 \cdot H_2O$	$CaAl_2Si_7O_{18} \cdot 6H_2O$	$(K, Na)_{0-1}(Ca, Na, Fe, Mg)_2$ $(Mg, Fe, Al)_5 (Si, Al)_8 O_{22}(OH)_2$
Colour	White, greenish, bluish*	Colourless, white, yellow, red*	Dark green to black*
Streak	White*	White*	White
Lustre	Vitreous	Vitreous to pearly*	Vitreous, silky*
Structure and form	mm2, prismatic, massive, granular, fibrous, stalactitic, mammilary	$2/m$, crystalline, platy diamond shaped*	$2/m$, prismatic, columnar*, fibrous*, bladed*, massive

(Contd...)

Cleavage	1 set perfect	1 set perfect	2 sets perfect at 56°*
Fracture	Uneven	Subconchoidal	Uneven
Hardness	4.5–5*	3.5–4*	5–6
Specific gravity	3.45, moderate	2.15, low	3–3.5, moderate
Special properties	Strongly pyroelectric and piezoelectric*, phosphorescent*, brittle, transparent to translucent	Transparent to translucent	Translucent
Occurrence	Secondary mineral in oxidised portion of zinc deposit.	It is a common zeolite mineral found in cracks or cavity walls in mafic igneous rocks and some metamorphic rocks.	Common in igneous rocks like granite, syenite and diorites and medium grade metamorphic rocks like amphibolites.
Use	Minor ore of zinc		

Mineral	Hypersthene	Ilmenite	Iron
Chemical composition	(Mg, Fe)$_2$Si$_2$O$_6$	FeTiO$_3$	Fe
Colour	Greenish black*, brownish green	Iron-black*	Steel-grey to black*
Streak	White, grey	Brownish red to black*	Black*
Lustre	Vitreous to pearly*	Metallic to submetallic*	Metallic*
Structure and form	2/m 2/m 2/m, prismatic*, crystalline, massive	$\bar{3}$, tabular, prismatic*, massive, granular, scaly	4/m $\bar{3}$ 2/m, blebs, plates, lamellar masses
Cleavage	2 sets perfect at 90°*	Absent	Absent
Fracture	Uneven	Subconchoidal	Hackly*
Hardness	5.5–6	5.5–6	4.5
Specific gravity	3.2–3.5, moderate	4.5–5, high*	7.3–7.9, very high*
Special properties	Translucent	Opaque, feebly magnetic*	Malleable, opaque, strongly magnetic*
Occurrence	Mafic igneous rocks and high-grade metamorphic rocks	Igneous rocks, pegmatites, high-grade metamorphic rock, black sands	Rare, common in meteorites
Use		Major source of titanium, which is used as white pigment; mixtures of ilmenite-magnetite and ilmenite-hematite are smelted for separation of Fe and Ti	Most useful metal now a days

(Contd...)

Mineral	Jadeite	Jamesonite	Jasper	Kaolinite
Chemical composition	$NaAlSi_2O_6$	$(4PbS \cdot FeS \cdot 3Sb_2S_3)$	SiO_2	$Al_2Si_2O_5(OH)_4$
Colour	White, shades of green*	Lead grey*	Red*, brown, yellow	White*
Streak	White	Greyish black*	White*	White
Lustre	Vitreous, greasy, pearly on cleavage surface*	Metallic	Dull*	Dull to earthy* in massive variety, pearly in crystals
Structure and form	2/m, fibrous, compact massive aggregates	2/m, acicular, fibrous	Granular, microcrystalline*	$\bar{1}$, massive, fine-grained aggregate, compact or friable clay like mass*
Cleavage	2 sets perfect	1 set distinct	Absent*	Imperfect
Fracture	Uneven	Uneven	Conchoidal to uneven	Conchoidal
Hardness	6.5–7	2–3*	7*	2–2.5*
Specific gravity	3.3, moderate	5.5–6, high	2.65, moderate	2.6, moderate
Special properties	Translucent	Opaque	Translucent	Translucent, earthy smell when blown by mouth*, plastic*
Occurrence	Metamorphic rocks of blueschist facies	Hydrothermal deposits	Banded iron ore	Formed by weathering of feldspar. Commonly associated with Gondwana sediments
Use	Manufacture of utensil, weapons and implements in stone age, ornamental stone	Minor source of lead and antimony	Decorative stone, implements in stone age	Manufacture of china-ware porcelain and tiles; fire bricks, plugs, pipes, funnels, adobe bricks, brick cement, etc. In paper production, kaolin is used as a filler and finishing agent. Manufacture of oilcloth, linoleum and pigments also in drilling for oil and salt.

(Contd...)

Mineral	Kernite	Kyanite	Lawsonite	Lazulite
Chemical composition	$Na_2B_4O_6(OH)_2 \cdot 3H_2O$	Al_2SiO_5	$CaAl_2Si_2O_7(OH)_2 \cdot H_2O$	$(Mg, Fe)\,Al_2(PO_4)_2(OH)_2$
Colour	Colourless or white	Blue*, white, grey	Colourless, pale blue to bluish grey	Azure-blue*
Streak	White	White	White	White*
Lustre	Vitreous to pearly*	Vitreous to pearly*	Vitreous to greasy*	Vitreous
Structure and form	2/m, prismatic, massive	$\bar{1}$, tabular, bladed*, radiating aggregates	222, tabular, bladed	2/m, prismatic, massive, granular
Cleavage	2 sets perfect	2 sets perfect	2 sets perfect	Imperfect
Fracture	Uneven	Uneven	Uneven	Uneven
Hardness	3*	5 parallel to length of crystal and 7 at right angles*	8*	5–5.5*
Specific gravity	1.95, low	3.6, high	3.1, moderate	3, moderate
Special properties	Transparent	Transparent to translucent	Translucent	Translucent
Occurrence	In association with borax deposits	Metamorphic mineral found in medium- to high-grade schists and gneisses	Metamorphic rocks of blueschist and glaucophane schist facies	Pegmatites and high-grade quartz rich metamorphic rocks
Use	A source of borax and boron compounds	Source of mullite for manufacture of refractories and acid resistant articles, spark plugs, crucibles, etc.		A minor gemstone

Mineral	Lazurite	Lepidocrocite	Lepidolite	Leucite
Chemical composition	$(Na, Ca)_8(Al, SiO_4)_6(SO_4, S, Cl)_2$	$(FeOOH)$	$K(Li, Al)_{2-3}(AlSi_3O_{10})(O, OH, F)_2$	$KAlSi_2O_6$
Colour	Azure-blue, greenish blue*	Red to black*	Lilac to rose-red*, yellow, grey, white	White, grey*
Streak	White*	Orange*	White	White
Lustre	Vitreous	Adamantine	Pearly*	Vitreous to dull
Structure and form	$\bar{4}3m$, crystalline, massive	2/m 2/m 2/m, platy, fibrous aggregate	2/m, six-sided prismatic*, coarse to fine-grained scaly* aggregate, micaceous*	4/m below 605°C; 4/m $\bar{3}$ 2/m above 605°C

(Contd....)

Property				
Cleavage	Imperfect	2 sets perfect	1 set perfect	Imperfect*
Fracture	Uneven	Uneven	Uneven	Conchoidal
Hardness	5–5.5	4*	2.5–4*	5.5–6*
Specific gravity	2.38–2.45, low	4–4.1	2.9, moderate	2.48, low
Special properties	Translucent, opaque*	Transparent	Translucent	Translucent to translucent
Occurrence	Crystalline limestones formed by contact metamorphism	Hydrothermal deposits	Li-rich pegmatites	Si-poor K-rich volcanic rock
Use	Lapis Lazuli (a mixture of lazurite and calcite) is an ornamental stone, previously used as paint and pigment		Source of lithium and lithium salts used in alkali storage batteries (in submarines), special optical glass, fireworks, medicine, artificial mineral water, air-conditioning, purification of helium, photography, etc.	For making potassium products and production of potash-fertilizer

Mineral	Limonite/hydrogoethite	Lithiophillite	Lollingite	Loparite
Chemical composition	$(HFeO_2 \cdot nH_2O)$	$Li\,(Mn, Fe)\,PO_4$	$FeAs_2$	$(Na, Ce, Ca)\,(Nb, Ti)O_3$
Colour	Brown to black*	Clove brown*	Silver-white to lead-grey*	Black to greyish black*
Streak	Yellowish-brown*	Greyish white	Greyish-black*	Brown*
Lustre	Submetallic, earthy	Vitreous to resinus*	Metallic	Submetallic
Structure and form	$2/m\ 2/m\ 2/m$, reniform, spongy*	$2/m\ 2/m\ 2/m$, massive, cleavable	$2/m\ 2/m\ 2/m$, prismatic, massive	$4/m\ \bar{3}\ 2/m$, cubic, massive
Cleavage	Absent	1 set perfect	1 set distinct	Absent
Fracture	Subconchoidal to uneven	Subconchoidal to uneven	Uneven	Uneven
Hardness	4.5–5.5*	4.5–5*	5–5.5	5.5–6
Specific gravity	3.6–4, high	3.42–3.56, moderate	7.0–7.4, very high	4.75–4.89, high
Special properties	Opaque	Translucent	Opaque, good conductor of electricity*	Translucent
Occurrence	As altered product with other iron minerals	Pegmatite	Hydrothermal vein and metasomatic deposits	Pegmatite dyke
Use	Minor source of iron	Source of lithium	Source of arsenic	Source of niobium, rare earths and titanium

(Contd...)

Mineral	Magnesite	Magnetite	Malachite	Manganite
Chemical composition	$MgCO_3$	Fe_3O_4	$Cu_2CO_3(OH)_2$	$MnO(OH)$
Colour	White, grey, brown, yellow	Black*	Bright green*	Grey to black*
Streak	White	Black*	Pale green*	Brown*
Lustre	Vitreous, porcelainous*	Metallic*	Adamantine to vitreous*	Metallic to submetallic*
Structure and form	$\bar{3}$ 2/m, earthy masses*	4/m $\bar{3}$ 2/m, octahedral, massive, granular	2/m, botryodal, colloform*, banded, massive	2/m, prismatic*, stalactitic, columnar, bladed, fibrous
Cleavage	3 sets perfect rhombohedral in crystals*	Absent	1 set perfect	2 sets perfect
Fracture	Conchoidal	Subconchoidal	Subconchoidal	Uneven
Hardness	3.5–5*	6	3.5–4*	4*
Specific gravity	3, moderate	5.2, high*	3.7–4, high*	4.2–4.4, high*
Special properties	Transparent to translucent, reacts with hot HCl*	Opaque, strongly magnetic*	Translucent, reacts with HCl*	Opaque
Occurrence	Veins and irregular masses derived from the alteration of Mg-rich igneous and metamorphic rocks	Accessory mineral of igneous, metamorphic and sedimentary rocks, concentrated to form ore bodies by magmatic, metamorphic or sedimentary processes	A secondary copper mineral found in the oxidized part of copper veins	Uncommon secondary mineral found in low temperature hydrothermal veins
Use	Production of CO_2, Mg and Mg-salts, refractory bricks, crucibles, special cement, used in paper and sugar industries	Ore of iron; the slag obtained from titanomagnetite ore is a source of vanadium	Ore of copper; as facing and mosaics, vases, trinket boxes and other ornamental objects; also used for manufacture of pigment	Source of manganese and important material for ferromanganese and other ferroalloys used in steel making

Mineral	Marcasite	Margarite	Microcline	Millerite
Chemical composition	FeS_2	$CaAl_2(Al_2Si_2O_{10})(OH)_2$	$KAlSi_3O_8$	NiS
Colour	Bronze-yellow* to greenish white	Pink*, white, grey	White, green* (Amazonite)	Brass yellow*
Streak	Greyish black*	White*	White	Greenish black*

(Contd...)

Property				
Lustre	Metallic*	Vitreous to pearly	Subvitreous	Metallic*
Structure and form	2/m 2/m 2/m, tabular, radiating*, needle like, coloform, globular, reniform, stalactic	2/m, massive, micaceous books*, disseminated particles	1̄, crystalline, prismatic	3̄ 2/m, acicular, filiform, radiating
Cleavage	Imperfect	1 set perfect	2 sets perfect*	1 set perfect
Fracture	Uneven	Uneven	Uneven	Uneven
Hardness	6–6.5	3.5–5*	6*	3–3.5*
Specific gravity	4.9, high*	3.1, moderate	2.56, moderate	5.5, high*
Special properties	Opaque	Transparent to translucent, brittle	Translucent	Opaque
Occurrence	Metalliferous veins with Pb and Zn ores, replacement mineral in limestone and shale	In association with corundum	Granites, pegmatites, syenites, sandstone and metamorphic rocks	Replacement of other Ni minerals and in cavities
Use	Minor source of sulphur, production of H_2SO_4, $Fe(SO_4)$, alum, etc.		Manufacture of glass and ceramics (porcelain); green amazonite is used as ornamental material	A minor ore of nickel

Mineral	Mimetesite (Mimetite)	Molybdenite	Monazite	Montmorillonite
Chemical composition	$Pb(AsO_4)_3Cl$	MoS_2	$(Ce, La, Y, Th) PO_4$	$(Ca, Na)_{0.2-0.4} (Al, Mg, Fe)_2 (Si, Al)_4O_{10} (OH)_2$
Colour	White, honey-yellow*, brown, green*	Lead-grey*	Yellowish to reddish brown*	White, grey
Streak	White	Greenish grey*, black	White*	White, grey
Lustre	Adamantine, greasy	Metallic*	Resinous*	Dull*
Structure and form	6/m, prismatic, acicular	6/m 2/m 2/m, foliated*, massive, scales	2/m, tabular, prismatic, granular mass, grains in sand	2/m, massive*
Cleavage	Imperfect	1 set perfect	1 set perfect	1 set perfect
Fracture	Conchoidal to uneven	Conchoidal	Subconchoidal	Irregular
Hardness	3.5*	1–1.5*	5–5.5	1–1.5*
Specific gravity	7.2–7.3	4.7, high*	4.9–5.2, high*	2–2.7, low
Special properties	Opaque	Opaque, sectile*, greasy feel*	Translucent	Translucent

(Contd...)

Occurrence	Oxidised zone of lead–zinc sulphide deposits	Granite pegmatite, porphyry copper deposit, vein deposit	Granite, pegmatites, gneisses, placer deposits	Residual, formed by alteration of Al-rich silicates
Use	Minor ore of lead	Chief source of molybdenum, which is used in high-grade steel, manufacture of dyes, paint, radio equipment, chemical industry	Chief source of thorium and rare earth elements	Decolorizing of oils, water softener, as a filler for thickening of drilling mud, for dyeing fabrics; in rubber industry as fillers; also used in production of soap, lipstick, toothpowder and paste; as additives in making paper and ceramics, as binder in medical preparations.

Mineral	Muscovite	Natrolite	Nepheline
Chemical composition	$KAl_2(AlSi_3O_{10})(OH)_2$	$Na_2Al_2Si_3O_{10} \cdot 2H_2O$	$(Na, K) AlSiO_4$
Colour	Colourless	Colourless or white	Colourless, white, yellowish*
Streak	White	White	White
Lustre	Vitreous, silky, pearly*	Vitreous	Vitreous in clear crystals, greasy in massive type*
Structure and form	2/m, foliated*, micaceous*, books	mm2, acicular, radiating crystal groups*, fibrous, massive, granular	6, prismatic, massive
Cleavage	1 set perfect	1 set perfect	1 set distinct
Fracture	Even	Uneven	Subconchoidal
Hardness	2–2.5*	5–5.5	5.5 – 6
Specific gravity	2.8, moderate	2.23, low	2.6, low
Special properties	Transparent* and translucent, elastic*, scratched by finger nail*	Transparent to translucent, brittle, soluble in HCl*	Transparent to translucent, soluble in HCl*
Occurrence	Granite, granite-pegmatite, mica-schists	Found lining the cavities in basalt with other zeolites	Si-poor igneous rocks like syenite
Use	Insulator in electrical industry, giving gloss to wallpapers, manufacture of lubricants, wall-finishes, rubber tyres, filler and fireproof material in furnace and stove doors; high explosives (as an absorbent).		Making of green glass and also in ceramic industry

(Contd...)

Mineral	Nicolite	Olivine	Opal
Chemical composition	NiAs	$(Mg, Fe)_2SiO_4$	$SiO_2 \cdot nH_2O$
Colour	Copper-red*	Yellowish green to brown*	White, pinkish white, grey
Streak	Brownish black*	White to yellowish white*	White
Lustre	Metallic*	Vitreous	Vitreous, resinous*
Structure and form	6/m 2/m 2/m, massive, reniform*, columnar	2/m 2/m 2/m, crystalline, lozenge shaped*, granular mass	Amorphous*
Cleavage	Imperfect	Imperfect	Absent*
Fracture	Uneven	Conchoidal	Conchoidal
Hardness	5–5.5	6.5–7	5–6
Specific gravity	4.6, high*	3.27–4.37, high	2–2.5, low
Special properties	Opaque	Transparent to translucent, slowly soluble in HCl	Transparent to translucent
Occurrence	Sulfide deposit hosted by mafic igneous rocks	Dark-coloured igneous rocks like gabbro, peridotite and basalt. Dunite is composed of only olivine	Deposited by hot springs, meteoric water and low temperature hypogene solutions, lining and filling cavities, siliceous tests of silica-secreting organisms.
Use	A minor ore of nickel	Manufacture of refractory brick; transparent green variety (peridot) is used as gemstone	Precious opals are used as ornamental stones; diatomite, a variant of opal is used as abrasive, filler, filtration powder and in insulation products; tripoli is used for polishing metals and stones.

Mineral	Orpiment	Orthoclase	Pectolite
Chemical composition	As_2S_3	$KAlSi_3O_8$	$Ca_2NaH(SiO_3)_3$
Colour	Lemon yellow to orange*	White, flesh-red*, grey	White, colourless, grey
Streak	Pale yellow*	White	White
Lustre	Resinous, pearly on cleavage*	Vitreous, pearly*	Vitreous to silky*
Structure and form	2/m, tabular, columnar, foliated*	2/m, prismatic*, elongate	$\bar{1}$, acicular, radiating, fibrous*
Cleavage	1 set perfect	2 sets perfect at 90°*	2 sets perfect
Fracture	Even	Conchoidal to uneven	Uneven

(Contd...)

Hardness	1.5 – 2*	6*	4.5–5*
Specific gravity	3.49, moderate	2.56, moderate	2.9, moderate
Special properties	Translucent, sectile*	Translucent	Transparent to translucent; decomposed by HCl
Occurrence	Hot spring, gold deposit	Silicic igneous rocks, arkose and metamorphic rocks	Lining cavities with zeolites
Use	Dyeing, removal of hair from skin	Manufacture of porcelain and pottery, opalescent glass; production of glazes on earthenware and enameled brick, sanitary ware; binder for abrasive wheels, mild abrasive, facing of artificial building materials. Opalescent to pearly variety of orthoclase (moon stone) is used as gem	

Mineral	Pentlandite	Periclase	Perovskite	Phenacite
Chemical composition	(Ni, Fe)$_9$S$_8$	MgO	CaTiO$_3$	(Be$_2$SiO$_4$)
Colour	Yellowish bronze*	Colourless, grey	Greyish black, brown, yellow*	Colourless, yellow, pink, brown*
Streak	Light bronze brown*	Orange-yellow*	White or greyish yellow*	White
Lustre	Metallic*	Vitreous	Adamantine	Vitreous, greasy
Structure and form	4/m $\overline{3}$ 2/m, massive, granular aggregates	4/m $\overline{3}$ 2/m, crystalline, granular mass	4/m $\overline{3}$ 2/m, cubic, reniform	$\overline{3}$, rhombohedral, short-columnar
Cleavage	2 sets perfect	2 sets perfect*	3 sets distinct*	Imperfect
Fracture	Uneven	Uneven	Uneven	Conchoidal
Hardness	3.5–4*	5.5	5.5–6	7.5–8*
Specific gravity	5, high*	3.56, high	3.97, high	2.96–3.0, high
Special properties	Opaque, brittle	Transparent to translucent	Transparent to translucent	Transparent to translucent
Occurrence	Basic igneous rocks, sulphide deposits	High temperature metamorphic carbonate contact aureoles	Contact metasomatic deposits	Pegmatite vein
Use	Source of Ni; Co, Cu, Pt metals are obtained as byproduct			Source of beryllium; transparent coloured varieties are used as gemstone

(Contd...)

Mineral	Phillipsite	Phlogopite	Pigeonite	Plagioclase
Chemical composition	$(K_2, Ca)(Al_2Si_4O_{12}) \cdot 4.5H_2O$	$KMg_3(Al, Si_3O_{10})(OH)_2$	$(Ca, Mg, Fe)_2Si_2O_6$	Albite: $NaAlSi_3O_8$ Anorthite: $CaAl_2Si_2O_8$
Colour	White, red*	Yellowish-brown*, green, white	Brown, green, black*	Colourless, white*, grey
Streak	Colourless*	White	White	White
Lustre	Vitreous	Vitreous to pearly*	Vitreous	Vitreous to pearly*
Structure and form	2/m, columnar, massive	2/m, flakes*, foliated masses	2/m, prismatic, columnar*	$\bar{1}$, crystalline, cleavable masses*
Cleavage	2 sets distinct	1 set perfect	2 sets perfect	2 sets perfect*
Fracture	Uneven	Even	Uneven	Conchoidal to uneven
Hardness	4–4.5*	2.5–3*	6	6*
Specific gravity	2.2	2.8, moderate	3.2–3.5, moderate	2.62, moderate
Special properties	Translucent to opaque	Transparent, flexible and elastic*, decomposed in H_2SO_4	Transparent to translucent	Transparent to translucent, some varieties show play of colours* and opalescence*
Occurrence	Cavities in basalts and phonolites	Marbles, Mg-rich igneous rocks, ultrabasic rocks	High temperature rapidly cooled lavas, diabase intrusives	Common in igneous and metamorphic rocks, rare in sedimentary rocks
Use		Insulator in electrical industry, fire-proof material, furnace and stove doors; heat-resistant paints, ceramics, tyres, lubricant, etc.		Albite is used in glass and ceramic industry, labradorite showing play of colours is used as an ornamental stone, varieties showing opalescence are known as moonstone and are used as gemstone

Mineral	Platinum	Polybasite	Powellite
Chemical composition	Pt	$(Ag, Cu)_{16}Sb_2S_{11}$	$CaMoO_4$
Colour	Silver-grey*, steel-grey*	Iron-black	Yellow, yellowish-green*
Streak	Silver-grey*, steel-grey*	Black*	Yellowish to greenish*
Lustre	Metallic*	Metallic*	Adamantine*
Structure and form	4/m $\bar{3}$ 2/m, irregular masses, grains, nuggets*	2/m, prismatic, tabular	4/m, tabular, foliated

(Contd...)

Cleavage	Absent	Imperfect	Absent
Fracture	Hackly*	Uneven	Uneven
Hardness	4–4.5*	2–3*	3.5*
Specific gravity	21.47, very high	6–6.33, very high	4.25 – 4.52
Special properties	Opaque, malleable*, ductile	Opaque, sectile	Opaque, brittle
Occurrence	In ultramafic rocks like dunite, placer deposits	Silver containing hydrothermal vein	Oxidized zone of molybdenum deposit
Use	As a catalyst in chemical and petroleum industry, chemical apparatus, electrical equipment, jewelry, dentistry, surgical instruments, pyrometry and photography, manufacture of platinum crucible.	Minor ore of silver and antimony	Secondary source of molybdenum and tungsten

Mineral	Prehnite	Psilomelane	Pyrargyrite – Proustite
Chemical composition	$Ca_2Al(AlSi_3O_{10})(OH)_2$	$BaMn_9O_{16}(OH)_4$	Pyrargyrite: Ag_3SbS_3 Proustite: Ag_3AsS_3
Colour	Pale green to white*	Black*	Pyrargyrite: Red* Proustite: Vermilion*
Streak	White	Brownish black*	Pyrargyrite: Red* Proustite: Vermilion*
Lustre	Vitreous	Submetallic, dull	Adamantine*
Structure and form	mm2, tabular, botryoidal*, globular, reniform*, stalactitic	2/m, massive, botryoidal*, stalactitic, reniform*	3m, massive, disseminated grains
Cleavage	1 set perfect	Absent	1 set perfect
Fracture	Uneven	Subconchoidal to uneven	Subconchoidal
Hardness	6–7*	5–6*	2–3*
Specific gravity	2.9, moderate	3.5–4.7, high*	Pyrargyrite: 5.85, high* Proustite: 5.57, high*
Special properties	Transparent to translucent	Opaque	Translucent
Occurrence	Secondary mineral lining cavities in basalt and related rocks, low-grade metamorphic rock	Secondary mineral associated with other Mn minerals	Low temperature silver veins and in replacement deposits
Use		Source of manganese and important material for ferromanganese and other ferroalloys used in steel making	Ores of silver

Mineral	Pyrochlore	Pyrite	Pyrolusite
Chemical composition	$(Na, Ca)_2 (Nb, Ti)_2O_6 (F, OH)$	FeS_2	MnO_2
Colour	Brown*	Brass-yellow*	Black*
Streak	Brown*	Greenish to brownish black*	Black*
Lustre	Vitreous, resinous*	Metallic*	Metallic, dull*
Structure and form	$4/m \bar{3} 2/m$, octahedral, colloform, massive	$2/m \bar{3}$, striated cube, pyritohedron*, massive, granular, reniform, globular, stalactitic	$4/m 2/m 2/m$, radiating*, fibrous, reniform, columnar, dendritic (crystalline variety—polianite)
Cleavage	Absent	Imperfect	Imperfect
Fracture	Conchoidal	Subconchoidal	Uneven
Hardness	5–5.5	6–6.5	1–2*, Polianite – 6
Specific gravity	4.0–4.4	5.1, high*	4.5–5, high
Special properties	Subtranslucent to opaque	Opaque, brittle	Opaque, soils hand*, marks paper*, brittle
Occurrence	Pegmatites in association of nepheline-syenite	Most common sulphide mineral, accessory mineral in igneous rock, contact metamorphic deposits, hydrothermal veins and sedimentary rock	Secondary mineral found as coatings, nodules, dendrites and as in bed. (Wad is the Mn-ore composed of impure mixture of hydrous Mn-oxides)
Use	Source of niobium, tantalum and uranium	Mined for gold and copper associated with it. Ore of iron (where oxide ores not available). Chief source of sulfur, used in dyeing, manufacture of inks; production of H_2SO_4, Fe (SO_4), alum, etc., preservative of wood and as a disinfectant; In some cases Cu, Zn, Au, Se and Co are obtained as by-product.	Ore of Mn and also used for making dry battery; decolouriser of green glass; manufacture of chemicals for medicinal and other purposes; manufacture of special gas masks against carbon monoxide; drying of oils and wax; as a tanning agent for chrome leather; in photography; preparation of pigments.

Mineral	Pyromorphite	Pyrophyllite	Pyroxene	Pyrrhotite
Chemical composition	$Pb_5 (PO_4)_3Cl$	$Al_2Si_4O_{10}(OH)_2$	Ca, Mg, Fe silicate	$Fe_{1-x}S$
Colour	Green, yellow, brown*	White, grey, brown*	Augite—black* Diopside—green*	Brownish bronze*
Streak	White, yellow*	White	White	Grey to black*
Lustre	Resinous to adamantine*	Pearly to greasy*	Vitreous	Metallic*

(Contd...)

Structure and form	6/m, barrel shaped, cavernous, globular, reniform, fibrous, granular	2/m, massive and foliated, platy radiating masses	Crystalline, prismatic*, massive	$\bar{6}$ 2m above 300°C, 2/m below 300°C; hexagonal plates, massive, disseminated
Cleavage	Imperfect	1 set perfect	2 sets perfect*	Imperfect
Fracture	Uneven	Subconchoidal to uneven	Uneven	Uneven
Hardness	3.5–4*	1–2*	5–6	4*
Specific gravity	7, very high*	2.8, moderate	3.1–3.6, moderate	4.6, high*
Special properties	Subtransparent to translucent	Translucent, greasy feel*	Transparent to translucent	Opaque, magnetic*
Occurrence	Secondary mineral found in oxidized portion of lead veins	Low- and medium-grade metamorphosed shales	Basic and ultrabasic igneous rocks (pyroxinite), metamorphic rocks	Basic rocks like norite, contact metamorphic deposits, veins, pegmatites
Use	Source of lead	Used in manufacture of paper, ceramics, refractories, talcum powder, building stone insulators, rubber (as filler) and making idols, statues, etc.		Manufacture of sulphuric acid; in some cases, nickel, copper and platinum are obtained as byproduct

Mineral	Quartz
Chemical composition	SiO_2
Colour	Colourless or white, differently coloured depending upon the presence of impurities*
Streak	White
Lustre	Vitreous
Structure and form	32, 622, crystalline
Cleavage	Absent*
Fracture	Conchoidal
Hardness	7*
Specific gravity	2.65, moderate
Special properties	Transparent to translucent, piezoelectric
Occurrence	Essential and common ingredient of many igneous, metamorphic and sedimentary rocks

(Contd...)

Use: Transparent and beautifully-coloured varieties are used as gemstones. Colourless rock crystal is used for manufacture of optical instruments, quartz-wedge. Manufacture of piezoelectric crystal plates used in radio apparatus and circuits, oscillating devices (quartz clock). It is used in making highly refractory and acid-resistant laboratory utensils; manufacture of quartz lamps used in medicine as a source of ultraviolet radiation. Quartz sands are used in manufacture of glass and porcelain; Silicon carbide or carborundum (SiC), which is harder than orundum; ferrosilicon. Fine-grained quartz sands are used in sandblasters for polishing metal and stone articles, for cutting rocks, etc. Sandstone and quartzite are used as building material.

Mineral	Realgar	Rhodochrosite	Rhodonite
Chemical composition	AsS	$MnCO_3$	$MnSiO_3$
Colour	Red to orange*	Rose red*, light pink* to dark brown	Rose red*, pink*, brown
Streak	Red to orange*	White*	White
Lustre	Resinous*	Vitreous, pearly*	Vitreous
Structure and form	2/m, prismatic, massive, granular	$\bar{3}$2/m, rhombohedra, granular, colloform, globular, fibrous	$\bar{1}$, tabular, discrete grains, cleavable masses
Cleavage	1 set good	3 sets perfect*	2 sets perfect
Fracture	Conchoidal	Subconchoidal to uneven	Conchoidal
Hardness	1.5–2*	3.5–4*	5.5–6
Specific gravity	3.56, moderate	3.5–3.7, high	3.5–3.7, high
Special properties	Transparent to translucent, soils finger*, sectile*, garlic odour*	Transparent to translucent, soluble in hot HCl*	Transparent to translucent
Occurrence	Veins, volcanic sublimation product	Hydrothermal veins, Mn-rich metamorphic rocks, secondary mineral in residual deposits	Mn-deposits, Mn rich iron formations
Use	Source of arsenic trioxide, dying, firework, glass making	Ferromanganese and chemical industries. At times it contains smaller amounts of lead and silver; ornamental stone	Ornamental stone, imparts violet colour to glass and for glazing stoneware; used for vases, writing sets, column facing, etc.

Mineral	Riebeckite	Rutile	Samarskite	Sanidine
Chemical composition	$Na_2Fe_3^{2+} Fe_3^{3+} Si_8O_{22}(OH)_2$	TiO_2	$(Y, Er)_4[(Nb, Ta)_2O_7]_3$	$(K, Na)AlSi_3O_8$
Colour	Blue to black*	Red, reddish brown*	Black*	White
Streak	White to light blue*	Light brown*	Reddish-brown*	White
Lustre	Vitreous	Adamantine* to submetallic	Resinous, splendent*	Vitreous

(Contd...)

Structure and form	2/m, acicular, fibrous, asbestiform	4/m 2/m 2/m, prismatic, stubby to acicular	2/m 2/m 2/m, massive	2/m, tabular, prismatic
Cleavage	2 sets perfect	2 sets, 1 set distinct	Imperfect	2 sets perfect
Fracture	Uneven	Subconchoidal	Conchoidal	Conchoidal
Hardness	5	6–6.5	5–6	6
Specific gravity	3.1–3.4, moderate	4.24, high*	5.6–5.8	2.56, low
Special properties	Transparent to translucent	Transparent to translucent	Translucent to opaque	Transparent to translucent
Occurrence	Igneous rocks such as granite, syenite, nepheline syenite, pegmatite	Granite, granite-pegmatite, gneiss, mica-schist, metamorphic limestone and dolomite	Pegmatite dykes	Extrusive igneous rock like rhyolite and trachyte
Use	Crocidolite is an asbestiform variety of riebeckite. Crosidolite when replaced by quartz with preservation of fibrous texture, the resulting variety is used in jewelry under the name Tiger's eye or Cat's eye.	A source of ferrotitanium for making steel of high impact strength; as a brown pigment for ceramics; as a radio-wave detector; manufacture of titanium white, titanium extraction, coating of welding rods	Minor source of niobium, tantalum, etc.	Manufacture of glass and ceramics (porcelain)

Mineral	Scapolite Marialite-Meionite	Scheelite	Serpentine Antigorite-Chrysotile
Chemical composition	$Na_4(AlSi_3O_8)_3 Cl–$ $Ca_4 (Al_2Si_2O_8)_3 (CO_3, SO_4)$	$CaWO_4$	$Mg_3Si_2O_5 (OH)_4$
Colour	White, grey, pale green, yellow, bluish, reddish*	Colourless, white, yellow, brown, green*	Green to greenish black*
Streak	White	White	White*
Lustre	Vitreous to pearly*	Subadamantine* to vitreous	Greasy*, subresinous, silky
Structure and form	4/m, prismatic*, massive, granular	4/m, massive, columnar, granular aggregates	2/m, antigorite—platy* Chrysotile—fibrous, asbestiform*
Cleavage	1 set perfect	Imperfect	Absent
Fracture	Conchoidal	Uneven	Uneven
Hardness	5–6	4.5–5*	4*
Specific gravity	2.55, low	6.11, high*	2.5–2.6, low

(Contd...)

Special properties	Transparent to translucent decomposed by HCl*	Transparent to translucent, fluorescence in ultraviolet radiation*	Translucent, decomposed by HCl*
Occurrence	Schists, gneisses, amphibolites and granulite facies of rocks	Granite pegmatite, contact metamorphic deposits and high-temperature hydrothermal veins	Secondary mineral in mafic and ultramafic igneous rock formed by alteration of olivine, pyroxene and amphibole
Use	Transparent varieties as gemstones	Chief source of tungsten and tungsten compounds	Chrysotile is the chief source of asbestos. Beautifully coloured varieties are used as facing stones, trinket boxes, ash-trays, writing sets. Refractory for steel industry.
Mineral	*Siderite*	*Sillimanite*	*Silver*
Chemical composition	$FeCO_3$	Al_2SiO_5	Ag
Colour	Light to dark brown*	White, pale green*, brown	Silver-white*
Streak	White	White	Silver-white*
Lustre	Vitreous, pearly*	Vitreous	Metallic*
Structure and form	$\bar{3}$ 2/m, rhombohedral*, colloform, globular, botryoidal, fibrous	2/m 2/m 2/m, long slender prisms*, needles*, (fibrous variety is known as fibrolite)	4/m $\bar{3}$ 2/m, crystalline, acicular, flakes, plates, scales, filiform*
Cleavage	3 sets perfect*	1 set perfect*	Absent
Fracture	Subconchoidal	Uneven	Hackly*
Hardness	3.5–4*	6–7	2–3*
Specific gravity	3.96, high*	3.23, moderate	10.1–10.5, very high*
Special properties	Translucent, reacts with hot HCl*	Transparent to translucent	Opaque, malleable*, ductile*
Occurrence	Veins, clay ironstone, coal measures	High-temperature metamorphosed argillaceous rock	Oxidized zones of ore deposits, hydrothermal deposit, placer deposits
Use	Iron ore in Great Britain and Australia	Source of mullite for manufacture of refractory bricks and acid resistant articles	Ore of silver, silverware, coinage, filigree work, manufacture of crucibles, silver plating, silver chemicals, photographic film emulsions, plating, brazing, alloys, tableware and electronic equipment

(Contd...)

Mineral	Smithsonite	Sodalite	Sperrylite	Sphalerite (Zinc blende)
Chemical composition	$ZnCO_3$	$Na_3Al_3Si_3O_{12} \cdot NaCl$	$PtAs_2$	ZnS
Colour	Colourless, white, green, grey, blue, pink (yellow variety is Turkey-fat ore)	Blue*, white	Tin-white*	Colourless, brown, orange, yellow, black* (red variety is known as ruby zinc)
Streak	White	White	Black*	White, brown or yellow*
Lustre	Pearly*, vitreous	Vitreous	Metallic*	Adamantine, submetallic to nonmetallic, resinous*
Structure and form	$\bar{3}$ 2/m, rhombohedral, reniform, botryoidal, stalactitic, (honeycomb mass is known as dry-bone ore)	$\bar{4}$ 3 m, crystalline, massive	$2/m\,\bar{3}$, cubic, pyritohedral	$\bar{4}$ 3 m, crystalline, cryptocrystalline, botryoidal, cleavable mass
Cleavage	3 set perfect*	Imperfect	3 sets	6 sets perfect*
Fracture	Subconchoidal	Conchoidal	Subconchoidal	Conchoidal
Hardness	4–4.5*	5.5–6*	6–7	3.5–4*
Specific gravity	4.43, high*	2.3, low	10.5–10.7, very high*	4, high*
Special properties	Transparent to translucent, soluble in cold HCl*	Transparent to translucent, soluble in HCl	Opaque	Transparent to opaque, reacts with HCl*
Occurrence	Secondary mineral found with Zn-deposits in limestone	Si-deficient rocks	Pegmatites of basic magmas	Hydrothermal veins, replacement deposit
Use	Ore of zinc, ornamental stone		Chief source of platinum group of elements	Ore of zinc, important source of cadmium, indium, gallium, germanium; manufacture of brass, bronze and other alloys, electroplating, galvanization of iron, making zinc white, fluorescent screens

Mineral	Sphene (Titanite)	Spinel	Spondumen
Chemical composition	$CaTiSiO_5$	$MgAl_2O_4$	$LiAlSi_2O_6$
Colour	Grey, black, brown, greenish*	White, blue, green, brown, black*, red	White, grey, pink, green, yellow, purple*
Streak	White	White	White

(Contd....)

Lustre	Resinous* to adamantine*	Vitreous	Vitreous to pearly*
Structure and form	2/m, tabular, wedge-shaped*, lamellar, massive	$4/m\,\bar{3}\,2/m$, octahedral*, massive, irregular grains	2/m, prismatic with vertical striations*, cleavable mass*
Cleavage	1 set perfect	Imperfect	2 set perfect
Fracture	Uneven	Conchoidal	Uneven
Hardness	5–5.5	8*	6.5–7*
Specific gravity	3.5, moderate	3.5–4, high	3.15, moderate
Special properties	Transparent to translucent	Transparent to translucent	Transparent to translucent
Occurrence	Acid to intermediate igneous rocks and metamorphic rocks	Metamorphosed carbonates or schists, dark igneous rocks, placers	Granite pegmatite rich in lithium
Use	Source of TiO_2 for use as paint and pigment; a minor gemstone	Transparent and finely coloured varieties are used as gemstone. Ruby-Spinel—clear red, Spinel-Ruby—deep red, Balas-Ruby—rose red and Rubicelle—yellow.	Source of lithium for medical preparations, fireworks, photography, glass making, X-ray photography. Transparent coloured varieties are used as gemstone: Hiddenite—emerald green, Kunzite— pink, Triphane—colourless and yellow

Mineral	Stannite	Staurolite	Stephanite	Stibnite (Antimonite)
Chemical composition	Cu_2FeSnS_4	$Fe_2Al_9Si_4O_{23}(OH)$	Ag_5SbS_4	Sb_2S_3
Colour	Steel-grey with olive green tint*	Red brown to brownish black*	Greyish-black*	Lead-grey to black*
Streak	Black*	White to grey	Black*	Lead-grey to black*
Lustre	Metallic	Vitreous, resinous*	Metallic	Metallic*, splendent on cleavage surface
Structure and form	$\bar{4}\,2/m$, granular, massive	2/m, prismatic*, penetration twin* (cruciform cross)	2/m 2/m 2/m, prismatic, tabular	2/m 2/m 2/m, prismatic striated crystals*, slender, long, granular, acicular
Cleavage	Imperfect	Imperfect	1 set distinct	1 set perfect
Fracture	Uneven	Subconchiodal	Subconchiodal to uneven	Subconchiodal
Hardness	3–4*	7–7.5*	2–2.5*	2*
Specific gravity	4.3–4.5	3.75, high	6.2–6.3, high	4.6, high
Special properties	Opaque	Translucent	Opaque	Opaque

(Contd...)

Occurrence	Hydrothermal tin ore deposits	Medium and high-grade metamorphic rocks	Hydrothermal veins	Low temperature hydrothermal vein, replacement deposits, hot spring deposits
Use	Minor ore of tin	Transparent varieties are used as gemstones	Important ore of silver	Chief ore of antimony; also used in fireworks, matches, percussion caps, vulcanizing rubber, textile industries, glass manufacture and medicine

Mineral	Stilbite	Strontianite	Sulphur
Chemical composition	$CaAl_2Si_7O_{18} \cdot 7H_2O$	$SrCO_3$	S
Colour	White, grey, yellow, brown, red*	White, green, yellow, grey*	Bright yellow*
Streak	Grey	White	Yellowish white*
Lustre	Vitreous, pearly on cleavage*	Vitreous to resinous*	Resinous*
Structure and form	2/m, aggregates of twin crystals like sheaves of grain are common, cruciform twinned	2/m 2/m 2/m, prismatic*, acicular, fibrous, granular, massive, lamellar, frequently twinned giving pseudohexagonal habit	2/m 2/m 2/m, massive, colloform, stalactitic, reniform*
Cleavage	1 set perfect	2 sets perfect	Imperfect
Fracture	Subconchoidal	Uneven	Conchoidal to uneven
Hardness	3.5–4*	3.5–4*	1.5–2.5*
Specific gravity	2.15, low	3.72, high*	2.1, low
Special properties	Transparent to translucent, brittle, decomposed by HCl*	Transparent to translucent, reacts with HCl*	Transparent to translucent, sulphur smell*
Occurrence	Cavities in basalts and related rocks with other zeolites and schists associated with hydrothermal ore bodies	Hydrothermal veins	Associated with volcanic fumaroles, sulphide veins, sediments precipitated by bacterial activity
Use		Minor ore of strontium and preparation of strontium salts, which are used in manufacture of pyrotechnics, fireworks, ceramics, plastics, paints and purification of molasses.	Manufacture of H_2SO_4, fertilizers, insecticide, explosives, gunpowder, fireworks, coal-tar products, rubber; preparation of wood pulp for paper manufacture and making matches.

(Contd...)

Mineral	Sylvite	Talc
Chemical composition	KCl	$Mg_3Si_4O_{10}(OH)_2$
Colour	Colourless, white, shades of yellow, blue, red	Grey, white (compact and massive variety is known as steatite or soapstone)
Streak	White	White
Lustre	Vitreous	Resinous, silky, pearly to greasy*
Structure and form	$4/m\,\bar{3}\,2/m$, cubes*, massive, granular	$2/m$, foliated*, massive
Cleavage	Perfect cubic*	1 set perfect
Fracture	Uneven	Uneven
Hardness	2*	1*
Specific gravity	1.99, low*	2.8, moderate
Special properties	Transparent to translucent, salty and bitter taste*, scratched by finger nail*, soluble in water* and HNO_3*	Translucent, soapy feel*, sectile*, flexible, non-elastic, scratched by finger nail*, mark on cloth*
Occurrence	Evaporite deposits, salt flats, salt domes, sedimentary beds, deposits from volcanic gases	Low-grade metamorphic rocks like talc-schist; formed by alteration of Mg-silicates
Use	Mostly used in fertiliser industry; Compounds like KOH, K_2CO_3, KNO_3, $KClO_3$, $KMnO_4$, KCN, KBr, KI and others are used in medicine, photography, perfume, firework, paper, glass, varnish and paint industries	Manufacture of talcum and face powders; fireproof and unfading paints, soft pencils for marking glass, fabrics and metals; in paper and rubber industries as a filler, textile industry for bleaching cotton, removing grease spots, in electrical industry to make highvoltage insulators, also for glazes, acid- and alkali-proof vessels, rain water piping. Used as bricks and slabs; for lining furnaces and making idols, statues, etc.

Mineral	Tantalite-columbite	Tenorite	Tetrahedrite-tennantite	Tetradymite	Thorite
Chemical composition	$(Fe, Mn)(Nb, Ta)_2O_6$	CuO	Tetrahedrite—$Cu_{12}Sb_4S_{13}$ Tennantite—$Cu_{12}As_4S_{13}$	Bi_2Te_2S	$ThSiO_4$
Colour	Iron-black to brown*	Grey-black to black*	Greyish-black to black*	Steel-grey	Black, orange yellow*
Streak	Brown, dark red to black*	Grey-black*	Black to brown*	Grey	Brown to orange*
Lustre	Submetallic*	Submetallic*	Metallic to submetallic*	Metallic	Vitreous, greasy*

(Contd...)

Structure and form	2/m 2/m 2/m, prismatic, tabular, heart-shaped twins	1, scaly or earthy aggregate	$\bar{4}$ 3 m, crystalline, tetrahedron, dodecahedron and cube forms are common, massive, granular	$\bar{3}$, rhombohedral, granular, massive	4/m 2/m 2/m, granular, massive
Cleavage	1 set good	Imperfect	Imperfect	1 set perfect	Absent
Fracture	Subconchoidal	Subconchoidal to uneven	Uneven, subconchoidal	Subconchoidal to uneven	Conchoidal
Hardness	6	3–4*	3–4.5*	1.5–2*	4.5–5*
Specific gravity	5.2–7.9, high*	5.8–6.5, high	4.6–5.1, high*	7.2–7.6, high	4–5.4
Special properties	Subtranslucent, iridescent*	Opaque	Opaque, massive varieties are brittle	Opaque	Translucent
Occurrence	Granite pegmatite, carbonatite, placer deposits	Oxidized zone of copper sulphide deposits	Hydrothermal veins of Cu, Ag, Pb and Zn, contact metamorphic deposits	Hydrothermal goldquartz veins	Pegmatite, contact aureoles
Use	Source of tantalum and niobium	Minor ore of copper	Minor source of copper and silver	Source of bismuth and tellurium	Source of thorium

Mineral	*Torbernite*	*Topaz*	*Tourmaline*
Chemical composition	$Cu(UO_2)_2(PO_4)_2 \cdot 12H_2O$	$Al_2SiO_4(F, OH)_2$	$(Na, Ca)\,(Fe, Mg, Al, Li)_3 Al_6$ $(BO_3)_3 Si_6 O_{18}(OH)_4$
Colour	Emerald-green*	Colourless, yellow*, pink, yellowish, bluish, greenish	Black* (shorl), brown* (elbaite, dravite), green* (verdelite), yellow, red-pink (rubellite), blue* (indicolite), white, colourless (achrolite)
Streak	Greenish*	White	White
Lustre	Vitreous, pearly*	Vitreous	Vitreous to resinous*
Structure and form	4/m 2/m 2/m, tabular, scaly mass	2/m 2/m 2/m, prismatic*, granular	3 m; vertically striated* prismatic, fine columnar*, massive, compact
Cleavage	1 set perfect	1 set perfect	Imperfect
Fracture	Subconchoidal to uneven	Subconchoidal	Conchoidal to subconchoidal
Hardness	2–2.5*	8*	7–7.5*
Specific gravity	3.2, high	3.5–3.6, high	2.9, moderate

(Contd...)

Special properties	Transparent to subtranslucent, strongly radioactive*	Transparent to translucent, brittle	Translucent, strongly pyroelectric* and piezoelectric*
Occurrence	Oxidized zones of pegmatites and hydrothermal veins	Accessory mineral in granite, rhyolite, pegmatite and in contact aureoles adjacent to silicic plutons	Granite pegmatite, granite
Use	Minor source uranium	Transparent coloured varieties are used as gemstone	Gemstone: Green (Brazilian emerald), red-pink (rubellite), dark blue (indicolite), Due to its piezoelectric property, it is used for frequency control in radio transmitters and pressure gauge to measure transient blast pressures. Produces polarized light by selective absorption.

Mineral	Tremolite	Tridymite	Triphylite-Lithiophilite	Triplite
Chemical composition	$Ca_2Mg_5Si_8O_{22}(OH)_2$	SiO_2	$Li(Fe, Mn)PO_4$	$(Mn, Fe)_2(PO_4)F$
Colour	White, green*	Colourless	Triphylite—Bluish grey*, Lithiophilite—Clove brown*	Brown*
Streak	White	White	White, grey	Yellowish-brown*
Lustre	Vitreous	Vitreous	Vitreous to resinous*	Resinous to adamantine*
Structure and form	2/m; Prismatic, radiating blades, fibrous*, columnar, asbestiform*	Low (α) tridymite — 2/m High (β) tridymite—6/m 2/m 2/m	2/m 2/m 2/m, massive	2/m, prismatic, acicular, massive
Cleavage	2 sets perfect at 56°*	Absent*	1 set perfect	Imperfect
Fracture	Uneven	Conchoidal	Subconchoidal	Conchoidal
Hardness	5–6	7*	4.5–5.5*	4–5.5*
Specific gravity	3–3.3, moderate	2.28, low	3.5–5.5, high	3.4–3.8
Special properties	Transparent to translucent	Transparent to translucent	Transparent to translucent	Subtranslucent to opaque
Occurrence	Metamorphosed impure carbonate	High temperature silicic igneous rock	Pegmatite	Acid pegmatite dykes, hydrothermal veins
Use	Fibrous varieties are used as asbestos	Glass and ceramic industry	Source of lithium	

(Contd...)

Mineral	Turquoise	Ulexite	Uraninite (Pitch blende)	Vanadinite
Chemical composition	$CuAl_6(PO_4)_4(OH)_8 \cdot 4H_2O$	$NaCaB_5O_6(OH)_6 \cdot 5H_2O$	UO_2	$Pb_5(VO_4)_3Cl$
Colour	Blue, green*	White*	Black*	Red, brown, yellow, orange*
Streak	White, green*	White*	Brownish black*	White, yellow*
Lustre	Resinous, waxy*	Silky*	Pitchy dull* to submetallic	Resinous to adamantine*
Structure and form	$\bar{1}$, reniform, massive, granular, stalactitic	$\bar{1}$, fibrous, rounded masses; massive, (closely packed fibers variety is known as television rock*)	$4/m\ \bar{3}\ 2/m$, cubes, octahedral, colloform, massive, botryoidal	$6/m$, prismatic, rounded, globular
Cleavage	1 set perfect	1 set perfect	Absent	Absent
Fracture	Subconchoidal	Uneven	Conchoidal	Subconchoidal
Hardness	6	1–2.5*	5.5*	3*
Specific gravity	2.7, low	1.96, low*	6.5–9.5, high*	6.9, high*
Special properties	Translucent, soluble in hot HCl	Transparent to translucent, sectile*	Opaque, radioactive*	Transparent to translucent
Occurrence	Associated with Al-rich volcanic rocks	In arid regions from evaporation of water	Granite, pegmatite, high-temperature hydrothermal vein	Oxidized part of lead deposit
Use	Gemstone (turquoise matrixi—sky blue)	A source of borax	Chief ore of uranium and minor source of radium	Source of vanadium and minor source of lead

Mineral	Vermiculite	Vesuvianite (idocrase)	Vivianite
Chemical composition	$(Mg, Fe^{+2}, Fe^{+3})_3 (Si, Al)_4O_{10}(OH)_2 \cdot 4H_2O$	$Ca_{10}(Mg, Fe)_2\ Al_4Si_9O_{34}(OH)_4$	$Fe_2(PO_4)_2 \cdot 8H_2O$
Colour	Yellow, brown, green*	Brown, yellow, green, blue, red*	Colourless to transparent, blue to green* when altered
Streak	Uncoloured*	White	Colourless, blue*, brown
Lustre	Greasy*	Vitreous to resinous*	Vitreous to pearly*
Structure and form	$2/m$, sheet like*, foliated*, book like*	$4/m\ 2/m\ 2/m$, prismatic, striated-columnar, fibrous, granular	$2/m$, prismatic, acicular, reniform, earthy mass
Cleavage	1 set perfect	Imperfect	1 set perfect
Fracture	Ragged*	Subconchoidal	Subconchoidal to uneven

(Contd...)

Hardness	1–1.5*	6.5	1.5–2*
Specific gravity	2.4–2.7	3.4, moderate	2.58–3, moderate
Special properties	Transparent to translucent, expands on heating	Transparent	Transparent to translucent, sectile
Occurrence	Weathered product of biotite, phlogopite	Contact aureoles associated with impure limestone	Phosphorous-rich sedimentary iron ore
Use	Heat-insulator for steam pipes, boilers and furnaces; sound-insulator in aircraft; manufacture of wall paper, lubricant, etc.	Green variety (californite) is used as gemstone	Source of blue pigment

Mineral	Wavellite	Willemite	Witherite	Wolframite
Chemical composition	$Al_3(PO_4)_2(OH)_3 \cdot 5H_2O$	Zn_2SiO_4	$BaCO_3$	$(Fe, Mn)WO_4$
Colour	White, brown, yellow, green*	White, yellow-green, flesh-red*	Colourless, white, grey	Ferberite: Black* Huebnerite: Brown*
Streak	White	White*	White	Black to brown*
Lustre	Vitreous	Vitreous to resinous*	Vitreous to resinous*	Submetallic to resinous*
Structure and form	2/m 2/m 2/m, radiating spherulitic and globular aggregates	$\bar{3}$, massive to granular	2/m 2/m 2/m, columnar, globular, botryoidal*, twinned*	2/m, bladed, prismatic, lamellar, columnar, massive
Cleavage	1 set perfect	2 sets imperfect	2 sets, 1 set perfect	1 set perfect
Fracture	Subconchoidal	Subconchoidal	Uneven	Uneven
Hardness	3.5–4*	5.5	3.5	4–4.5*
Specific gravity	2.36, low	3.9–4.2, high*	4.29, high*	7.25–7.6, very high*
Special properties	Translucent	Transparent to translucent	Transparent to translucent, reacts with HCl*	Opaque
Occurrence	Rock cavities and joint surfaces, low-grade aluminous rocks, phosphorite deposits	Crystalline limestone, oxidized zone of zinc deposits	Low temperature veins	Pegmatite, high-temperature quartz-vein
Use		Minor Zn-ore; used in fluorescent screens in cathode-ray tubes and other instruments.	Secondary source of barium and Ba-salts; pottery industry	Chief source of tungsten and tungsten compounds

(Contd...)

Mineral	Wollastonite	Wulfenite	Wurtzite	Xenotime
Chemical composition	$CaSiO_3$	$PbMoO_4$	ZnS	YPO_4
Colour	Colourless, white, grey	Yellow, orange, red, brown, green*	Colourless, brown, orange, yellow, black*	Brown, red, grey*
Streak	White	White*	White to brown*	Brown to reddish*
Lustre	Vitreous, pearly to silky on cleavage surface*	Vitreous to adamantine*	Adamantine*	Vitreous, waxy, greasy
Structure and form	$\bar{1}$, cleavable masses, fibrous, tabular, prismatic	4 or 4/m, square tablets, tabular, pyramidal	6 mm, pyramidal, tabular, fibrous, massive	4/m 2/m 2/m, prismatic, massive
Cleavage	2 sets perfect	1 set distinct	2 sets, 1 set perfect	1 set perfect
Fracture	Uneven	Subconchoidal	Subconchoidal to uneven	Uneven
Hardness	5–5.5	3*	3.5–4*	4–5*
Specific gravity	3.1, moderate	6.7–7, high*	4–4.1	4.5–4.6
Special properties	Transparent to translucent, decomposed by HCl	Transparent to translucent	Opaque	Translucent
Occurrence	High-grade marbles and other calcareous rocks	Oxidised part of Pb-deposits	Hydrothermal deposit in association with sphalerite	Granite and pegmatite
Use	Manufacture of tile and white rock wool	Minor source of lead and molybdenum	Minor source of zinc	Minor source of rare earth elements

Mineral	Zincite	Zinnwaldite	Zircon
Chemical composition	ZnO	$KLiFe^{+2}Al (Si_3AlO_{10}) (F,OH)_2$	$ZrSiO_4$
Colour	White in purest form, red to orange-yellow* with Mn impurity	Grey-brown, white, green*	Grey, green, red, brown*
Streak	Orange-yellow*	Uncoloured to white*	Colourless to white*
Lustre	Subadamantine*	Vitreous, pearly*	Adamantine*
Structure and form	6 mm, massive, platy, granular	2/m, tabular, sheet, scaly	4/m 2/m 2/m, square prism*, pyramid, rounded
Cleavage	1 set perfect	1 set perfect	Imperfect
Fracture	Subconchoidal	Even	Conchoidal
Hardness	4–4.5*	2–3*	7.5*
Specific gravity	5.4–5.7, high*	2.9–3.2, high	4.68, high*

(Contd...)

Special properties	Translucent, soluble in HCl*	Transparent* and translucent, elastic*, scratched by finger nail*	Transparent to translucent
Occurrence	Zn-deposits	Li-rich pegmatites	Accessory mineral in igneous rock, particularly granite
Use	Ore of zinc; used for production of zinc white (ZnO)	Source of lithium and lithium salts used in alkali storage batteries (in submarines), special optical glass, fireworks, medicine, artificial mineral water, air-conditioning, purification of helium, photography, etc.	Source of zircon and ZrO_2, which are used in manufacture of acid resistant refractory crucible, refractory bricks, laboratory utensils, white enamel, scientific instruments, spark plugs, thermo-elements in pyrometers. Transparent coloured varieties are used as gemstone.

6

Optical Mineralogy

The minerals can be effectively and accurately identified by means of their optical properties, which are best studied under a microscope. For this method of study, the rock forming transparent minerals are cut into thin slices and the surfaces of opaque ore minerals are polished to produce glazed surfaces. Petrological microscope is necessary to study the optical characters of minerals. This microscope uses polarised light that vibrates in one direction unlike ordinary light that vibrates in all directions. Polarized light can be produced by selective absorption, double refraction and reflection. Most of the polaroids used in the petrological microscope now a days are made up of synthetic materials, which produce polarized light by selective absorption of one of the rays (ordinary ray). On the basis of their behaviour with light, the minerals are broadly divided into two groups, viz. isotropic and anisotropic. The isotropic minerals are single refracting and the light passing through them is not resolved into two rays. Minerals crystallizing in isometric (cubic) system belong to this category. On the other hand, the minerals crystallizing in other systems are double-refracting, as a result of which, the light passing through them is resolved into ordinary and extraordinary rays, both of which are plane polarised. The opaque minerals do not allow the light to pass through them. When the surface of an opaque mineral is polished and glazed, a part of light reflected from the surface is plane polarized.

6.1 PREPARATION OF THIN SECTION OF MINERALS

Thin sections of minerals are required for examination under microscope. A thin section is prepared in four stages, i.e. two stages of grinding and two stages of mounting.

a. First stage of grinding

The specimen is cut with a diamond saw. One side of a mineral chip is grinded and made perfectly flat with emery or carborundum powder on a metal plate that rotates by a motor. Grinding is made in several phases beginning with a coarse powder and continued with finer powders until a smooth flat surface is obtained. It is necessary to wash the chip to make it free from coarse powder at each stage of change of grinding powder.

b. First stage of mounting

A rectangular glass slide is taken, a small amount of Canada balsam is put on its center and heated gently. The exact stage to stop heating is ascertained by taking a small quantity of balsam at the tip of a forcep and making a small thread of balsam by opening the tip. At the exact stage of heating, the thread will be hard and brittle when cool. The mineral chip is placed with the flattened side in contact with balsam and glass. Air bubbles from the thin layer of balsam are removed by pressing the mineral chip against the glass slide. When cool, the mineral chip will be firmly fixed on the slide.

c. Second stage of grinding

The thick chip is grinded using coarse to fine emery or carborundum powder as was done earlier. Extreme care is taken in the final stage of grinding so that the mineral chip is not completely rubbed away. In the final phase of grinding, when the section is appreciably thin, grinding is done manually on a glass plate using finest powder. Intermittently, the mineral slice is viewed under microscope. These two operations (grinding and examination under microscope) are continued till the mineral section is 0.03 mm in thickness and shows its characteristic interference colour, if anisotropic. After final grinding, the section is carefully washed to remove the powder and the remaining balsam surrounding the mineral section is scrapped off.

d. Second stage of mounting

The mineral slice is covered with a drop of fresh Canada balsam and heated to a lesser extent than before. When the balsam is of right consistency, a thin sheet of glass known as cover slip is carefully placed over the mineral slice and pressed down with rotating motion so that no air bubble remains below the cover slip. The upper surface and surrounding areas of the cover slip are cleaned by a piece of cotton moistened with xylene, methylene alcohol or kerosene. Now the mineral section is ready for examination under microscope. Thin section of rock is prepared in exactly similar way. However, weathered rocks in which the constituent minerals or grains are loose are to be cooked (heated with sufficient Canada balsam in a crucible) prior to first phase of grinding.

6.2 PREPARATION OF POLISHED SECTIONS

Preparation of polished sections involves three steps. These are embedding the specimen, grinding and polishing.

a. First step: Embedding the specimen

The specimen is cut with a circular diamond saw and then put in a drying oven to eliminate all traces of water. Thereafter, the specimen is placed in a steel or plastic ring with an inner diameter of 4 cm resting on a glass plate. A thin film of silicon grease is placed between the plate and the ring to ensure a tight contact. An adequate quantity of Stratyl varnish with a small quantity of catalyst and hardener for polymerization, are poured on the ring and sample. After a few hours when the varnish is completely dry, the mounting is removed. If the sample is porous, superficial consolidation is undertaken with araldite, which polymerizes when hot with only one catalyst. For this, the specimen, which is already embedded in Stratyl, is impregnated with araldite and placed in a warming oven at 60 or 70 °C until it becomes solid.

b. Second step: Grinding

For this operation, either emery or carborundum powder is used. In the first case, emery with carborundum stuck on the disk of a semi-automatic machine is used. Abrasives with grain sizes of 120, 240, 400 and 600 mesh (ASTM) are used successively, at the rate of a few minutes per abrasive. Thereafter, the section is ground manually on a glass plate with carborundum powder of smaller grain sizes to the preceding ones. The first brief stage of grinding is followed by a second longer (about 1 hour) stage during which 1000 mesh carborundum (6 microns) is employed. Once this step is over, the mineral section is ready for polishing.

c. Third step: Polishing

Polishing is done either manually or by a set of semi-automatic machines. Very fine-textured aluminium papers are used for this purpose. Polishing time is as follows: Two periods of fifteen minutes with 6-micron diamond powder on aluminium paper; two periods of fifteen minutes with 3-micron diamond powder on aluminium paper; two periods of fifteen minutes with 1-micron diamond powder on aluminium paper; and a 30-minute final polish with ¼ micron diamond powder on a Lam Plan Matic cloth. After each polishing step, the section is properly cleaned. The diamond powders are kept in suspension in silicon oil mixed with a few drops of alcohol. In order to remove the last scratches, and to give the minerals some relief, so that their relative hardness can be studied, a very short polish generally follows the polishing stage with chromic oxide on soft winter felt. Alumina (corundum powder) is recommended for very soft minerals. Now the polished section is ready for observation under microscope.

6.3 PARTS AND FUNCTION OF PETROLOGICAL MICROSCOPES

The microscopes used to study the optical characters of rocks and minerals are different from biological and similar microscopes used to view small sized objects. These are known as petrological microscopes and use plane polarised light instead of ordinary light. Depending on the polarizing character of minerals, two types of petrological microscopes are used in the geological laboratories. These are refracting and reflecting types. In refracting microscope, light passing through the anisotropic mineral section is plane polarized. In case of reflecting microscopes, the light reflected from the mineral surface is plane polarized. However, nowadays both the facilities are available in one microscope. By selecting the mode (refracting/reflecting) the microscope can be used either as the refracting type to study the transparent mineral sections or as the reflecting type to study the polished opaque minerals.

6.3.1 Refracting Petrological Microscope

The refracting microscope shown in Fig. 6.1 stands on the base. A mirror is located at the lower part of the microscope. In many cases the mirror is concave type and reflects a concentrated beam of sunlight into the microscope. If the microscope is used at a place where sunlight cannot reach, fluorescent or ordinary tube light can be used as the light source. Many advanced microscopes are provided with self-illuminating light sources and are independent of sunlight. The filter, polariser, diaphragm and condenser are located between the mirror and the stage in case of refracting microscopes. These optical parts constitute the sub-stage. The filter is used to regulate the amount of light entering into the microscope.

The polariser splits the light into ordinary and extraordinary rays, both of which are plane polarized. The ordinary rays are cut off and the extraordinary rays move in upward direction from the polariser. These rays vibrate in left-right direction in case of synthetic polaroid or parallel to the shorter diagonal of the Nicol prism which is used as the polariser. The diaphragm regulates the amount of light that emerges out of the sub-stage and enters into the mineral section.

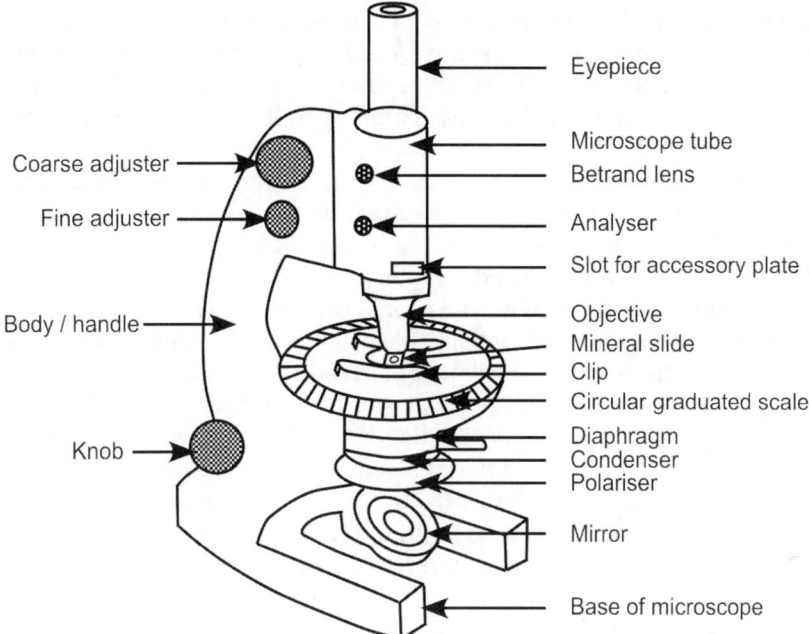

Fig. 6.1: Schematic diagram of a refracting type petrological microscope

The condenser is movable and is used for conoscopic illumination (Fig. 6.2). It causes the light beam to converge (focus) on a small spot of the thin section and illuminates the mineral with a cone of nonparallel rays. When the condenser is not in use, the illumination is said to be orthoscopic, i.e. perpendicular to thin mineral section (Fig. 6.2). The microscope stage is circular and graduated in degrees to make precise measurements of mineral orientation like angles between cleavages, crystal faces, twin orientations and other optical properties. The thin section of the mineral (or rock) is kept above the circular slit of the stage and is held in position by clips. The microscope stage can be rotated to change the orientation of the section relative to the polarized light. In case of anisotropic minerals the interaction of the light with the mineral varies with rotation of the stage. The polarized light leaving the substage passes through the transparent thin section of the mineral. If the mineral is isotropic, the light passes as such vibrating in left—right (or parallel to

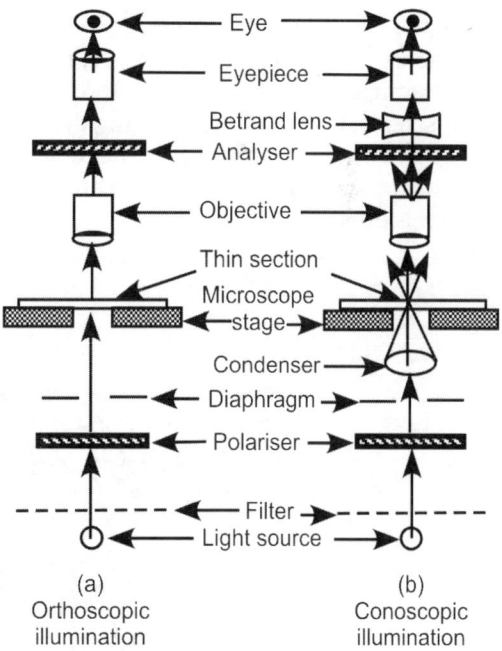

(a)
Orthoscopic
illumination

(b)
Conoscopic
illumination

Fig. 6.2: Orthoscopic and conoscopic illuminations in petrological microscope

shorter diagonal of the Nicol prism) direction. If the mineral is anisotropic and the light hits the mineral at an angle with the vibration directions, the light passing through the mineral is doubly refracted and resolved into ordinary and extraordinary rays, both of which are plane polarized and vibrate in directions perpendicular to each other (parallel to the vibration directions of the mineral). The microscope tube is equipped with an objective, slots for accessory plates, polariser, Betrand ocular and eyepiece. The ranges of magnification of objectives vary from 2× to 50×. The slots present in the NW and SE quadrants are aligned at 45° with the vibration directions of the polariser and analyser. Common   accessory   plates are gypsum plate, mica plates and quartz wedge. These are used for study of different properties of minerals in thin section. The analyser is removable. When it is not in use, the mineral is seen in plane polarized light. Grain shape, size, colour (pleochroism), cleavage, alteration, inclusion, refractive index are some of the properties studied in plane polarized light. The vibration direction of the analyzer is perpendicular to that of the polariser, i.e. north-south (top-bottom) in case of synthetic polaroid and parallel to the shorter diagonal of the Nicol prism. When the analyzer is in use, the mineral is said to be viewed in cross polars (crossed nicols, XP light). The ordinary and extraordinary rays leaving the anisotropic mineral undergo double refraction in the analyzer. Two ordinary rays are absorbed while two extraordinary rays vibrating in the same plane and having some phase difference interfere. Properties like istropism/anisotropism, interference colour, extinction, twinning, birefringence, etc. are studied under crossed nicols position. Like the analyzer, the Betrand lens is also removable. Sometimes, along with the cross polars and conoscopic illumination, the Betrand lens is used to determine properties like extinction position, retardation, optic sign and optic axial angle (2V).

6.3.2 Reflecting Petrological Microscope

In case of reflecting microscope (Fig. 6.3), the sub-stage consisting of filter, polariser, diaphragm, condenser with or without the light source are located between the objective and analyser. The polarised light leaving the sub-stage is reflected by the reflector, which is inclined at 45° with the microscope tube. The reflected light meets the polished section at right angle and is reflected back. If the mineral is anisotropic, the light reflected from the polished section surface is polarised.

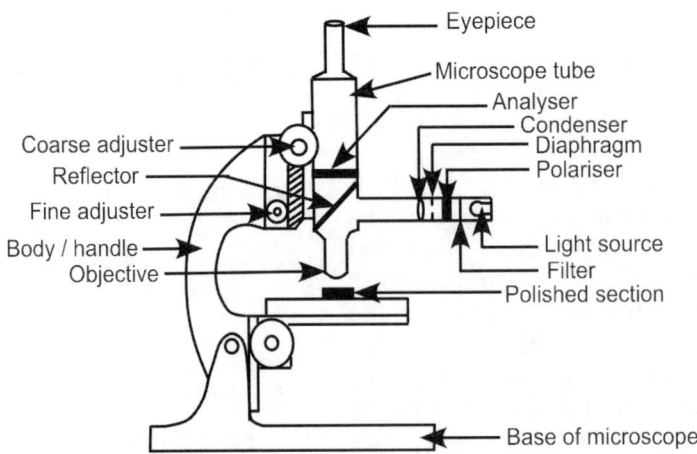

Fig. 6.3: Schematic diagram of reflecting type petrological microscope

Two rays (ordinary and extraordinary) emerging out of the anisotropic mineral while passing through the analyser are doubly refracted as a result of which two ordinary and two extraordinary rays are produced. The two ordinary rays suffer total internal reflection and are cut off. The two extraordinary rays while passing through the eyepiece vibrate in the same plane and possess some phase/path difference, as a result of which they interfere.

The magnification of the eyepiece varies from 4× to 10×. The total magnification produced by the microscopes varies from 8× to 500×. Some microscopes with greater magnifying power of objectives and eyepieces produce magnification more than 500×. The eyepiece is provided with cross wires aligned in north-south (top-bottom) and east-west (left-right) directions. These are used as reference directions for measurements. The eyepiece is removable and can be taken out when not required.

6.3.3 Production of Interference Colours

Constructive and destructive interference result depending upon the relationship of the phase difference (and thus path difference) between two interfering extraordinary rays with the wavelength (λ) of light used. If the phase difference is odd multiple of ($\frac{1}{2}$)λ, constructive interference takes place resulting in production of intense coloured light if monochromatic light is used. If the phase difference is even multiple of ($\frac{1}{2}$)λ, destructive interference takes place resulting in darkness. However, in almost all the microscopes white light is used as the illuminating source. The white light is a mixture of seven colours of different wavelengths. The wavelength of white light ranges from 380 to 760 mμ (1 mμ = 10^{-9} m = 10 Angstrom). Mean wavelengths of different constituent colors are 410 mμ—violet, 430 mμ—indigo, 460 mμ—blue, 510 mμ—green, 565 mμ—yellow, 620 mμ—orange and 690 mμ—red. The amount of phase/path difference between the interfering rays bears different relationship with different colours. As a result, some colours are suppressed and others are reinforced resulting in the production of one dominant colour or mixture of two or more colours (e.g. bluish-green, etc). The phase difference is directly related to the birefringence (difference of refractive indices of the mineral in different vibration directions) and thickness of the section. Birefringence, which is characteristic of a mineral, also depends on the direction of section with respect to crystallographic orientation. Both direction and thickness of the section depends on workmanship. The characteristic interference colours of different minerals mentioned in this book are shown by perfectly oriented sections of 0.03 mm in thickness. With this thickness, quartz shows straw yellow interference colour. But in thick rock sections quartz may be red, blue, green or differently coloured. With increasing phase difference, different interference (polarization) colours are produced. These are given in Table 6.1.

It is seen that with increasing path difference, same colour is repeated several times at interval of nearly 550 mμ. To distinguish the same colour from each other, the colours mentioned in Table 6.1 are grouped under different orders as shown in Table 6.2.

When the mineral is rotated, the interference colour decreases gradually and a position is reached when the mineral becomes completely black. This position is known as the *extinction position* and the mineral is said to *extinct*. It happens when the vibration directions of the mineral are parallel with the vibration directions of the analyser and polariser (i.e. the cross-wires). In a complete rotation of the stage (360°), the vibration directions of the mineral become parallel with the vibration directions of the analyser and polariser four times at intervals of 90°. Thus, an anisotropic mineral extincts four times in a complete rotation of 360°. At the positions, exactly

Table 6.1: Newton's scale of interference colours

Path difference (mμ)	Interference colour	Path difference (mμ)	Interference colour
0	Dark gray	850	Yellow
100	Gray	950	Orange
250	White	1050	Red
300	Yellow	1100	Violet
450	Orange	1250	Blue
530	Red	1350	Green
575	Violet	1450	Yellow
600	Blue	1500	Orange
750	Green	1650	Pink

Table 6.2: Order of interference colours

Order	Interference colour	Order	Interference colour
First	Dark gray Gray White Yellow Orange Red Violet	Third	Blue Green Yellow Orange Red Violet
Second	Blue Green Yellow Orange Red Violet	Fourth	Pale green, brownish pink and white

intermediate between the extinction positions, a mineral shows the maximum interference colour, which is characteristic of that mineral.

6.4 ACCESSORY PLATES

Each microscope is generally provided with a box containing extra objectives and eyepieces of different magnification, leveling screws and accessory plates. The leveling screws are used to center the field of view. There are three accessory plates, viz. mica and gypsum plates and quartz wedge.

6.4.1 Mica Plate

It is made up of a thin sheet of mica of such thickness that for yellow colour it gives a path difference of a quarter (¼) of a wavelength for which it is also known as *quarter wave plate*. The word *Glimmer* with (¼)λ is inscribed on it (Fig. 6.4a). It also carries a mark of *doubly terminated* arrow and 'γ or 'α' symbol that indicates the vibration direction of light in the plate. Normally the refractive indices for the fast and slow rays of biaxial minerals are indicated by symbols 'α' and 'γ' respectively. The arrow with "γ" indicates the direction of slow ray while arrow with "α" indicates the direction of fast ray. The mica plate shown in Fig. 6.4a is *fast along*, i.e. the fast (X) ray vibrates along the length of the plate while the slow ray (Z) vibrates perpendicular to the length. Mica plate is used to determine the fast and slow vibration directions and other optical characters of minerals.

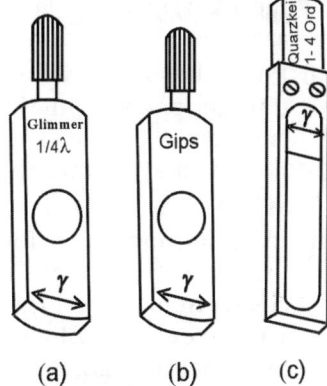

6.4.2 Gypsum Plate

It is made by cleaving a gypsum crystal of such a thickness that between crossed nicols (without mineral section) or when the mineral is in extinction position, it shows the violet at the end of first-order. Because, the human eye is sensitive to slight change in this colour, the plate is known as *sensitive tint plate*. Commonly the word *Gips* is engraved on this plate (Fig. 6.4b).

Fig. 6.4: Accessory plates; (a) Mica plate, (b) Gypsum plate and (c) Quartz wedge

Like the mica plate, from the arrow and 'α' or 'γ' symbols, the fast and slow vibration directions can be ascertained. It is used to know the exact extinction position in addition to determination of other optical characters of minerals. When superposed over a mineral, the gypsum plate gives blue when the phase difference is increased and yellow when phase difference is decreased.

6.4.3 Quartz Wedge

It is an elongated wedge-shaped piece of quartz (Fig. 6.4c). When it is inserted gradually into the slot in crossed nicol position without any mineral section, a series of interference colours starting from the first-order gray appear. Like the mica plate, from the arrow and 'α' or 'γ' symbols, the fast and slow vibration directions can be known. It is used to determine the order of interference colour, birefringence and optic sign of minerals.

In this book, the vibration directions of the accessory plates along their lengths have been taken as fast (X), as in case of many Lietz microscopes and the observations and properties of the minerals have been described accordingly. However, if the vibration directions of the accessory plates used in the laboratory are slow (Z) along their lengths, the observations and properties of the minerals will be exactly opposite. In case of some microscopes as in Kyowa of Japan make, the slot for accessory plates is aligned in NE–SW direction, i.e. perpendicular to that of Lietz microscopes. However, in these cases, the vibration directions parallel to length of accessory plates are slow (Z). So the minerals observed under Kyowa microscopes show the same properties as observed under Lietz microscopes with fast (X) vibration direction parallel with length of plates.

6.5 DETERMINATION OF SLOW AND FAST VIBRATION DIRECTIONS IN A MINERAL

The uniaxial and biaxial minerals have two and three vibration directions respectively, along which light travels with different velocities. While studying the optical characters, it is often required to determine the fast and slow vibration directions of light in the minerals. The procedure is outlined below.

i. Under crossed nicol, rotate the stage to bring the mineral to extinction position. At this position the vibration directions of the mineral are parallel with the vibration directions of the analyser and polariser (i.e. with cross-wires). The exact extinction position can be ascertained by gypsum plate. Note the stage reading.

ii. Rotate the stage by 45° either in clockwise or anticlockwise direction. In the former case the direction parallel to the horizontal cross-wire becomes parallel to slot (NW-SE), while in the later case, the direction parallel to the vertical cross-wire becomes parallel to slot. At this position the mineral will show the maximum interference colour. Note the colour. Say it is yellow.

iii. Insert the mica plate slowly and mark the change of interference colour. Let the original yellow colour changes to green. As shown in Table 6.2, green comes before yellow. So change from yellow to green is said to be a decrease of the interference colour. This happens when the natures of vibrations of the mineral and mica plate are opposite to each other. Since the mica plate is fast (X) along the length, the vibration direction of the mineral parallel to slot is slow (Z). In the opposite case, if the vibration direction of the mineral parallel to slot is fast (X), the yellow colour will change to red (as orange is difficult to recognize) indicating increase of the interference colour. In this case the natures of vibrations of the mineral and mica plate are equal with each other. In some cases, the increase and decrease of the interference colour is shown by a change of quality of colour, i.e. yellow decreases to light yellow or increases to deep yellow. Thus, the positions of fast (X) or slow (Z) vibration directions of light in the mineral are ascertained, which are perpendicular to each other, i.e. 90° apart.

6.6 OPTICAL PROPERTIES OF TRANSPARENT MINERALS IN THIN SECTION

The properties of the minerals in thin section can be studied both in plane polarised light and under crossed nicols.

6.6.1 Properties in Plane Polarised Light

In this case the polariser is in position and the analyser is not in use. In this case certain properties of the mineral like colour, pleochroism, pleochroic halos, refractive index, cleavage, shape and habit, alteration, inclusion and twinkling are studied.

6.6.1.1 Colour

In hand specimen many minerals appear coloured due to selective absorption of the visible spectrum. But in thin section or in grain mount they are not thick enough to cause appreciable amount of absorption or enhancement of specific wavelength of light. As a result, many minerals when viewed in plane polarised light display a weak colour or appear colourless. A few minerals like biotite, hornblende, chlorite, rutile, etc. show deep colours. Isotropic minerals (minerals crystallising in isometric system) do not exhibit any change in colour when rotated with the microscope stage. Some garnets are faint brown or yellow in plane-polarised light and the colour remains unchanged with any amount of rotation. This is due to the fact that in isotropic minerals the light rays vibrating in all directions have the same velocity and character. The anisotropic minerals are either uniaxial (crystallising in tetragonal and hexagonal systems) or biaxial (crystallising in orthorhombic, monoclinic and triclinic systems). In former case, there are two vibration directions (X and Z) while in later case there are three (X, Y and Z) vibration directions. Along these vibration directions light travels with different velocities and the selective absorption of the visible spectrum is also different. Thus, with the rotation of the mineral, different colours appear. This property of minerals is known as pleochroism.

6.6.1.2 Pleochroism

A mineral is said to be pleochroic when it shows change in quality or quantity of colour when rotated in plane polarised light. It is due to unequal absorption of light by the mineral in different directions. Biotite shows dark brown and pale yellow colour when the light vibrates parallel and perpendicular to cleavage respectively. The pleochroic formula for biotite is X = pale yellow, Y = Z = dark brown. The absorption formula is X > Y = Z. Similarly, hornblende shows yellow, bluish green and blue colours when light vibrates parallel to X, Y and Z directions respectively. The absorption formula is Z > Y > X or X < Y < Z. The method of determination of scheme of pleochroism is given below.

i. Determine the fast vibration direction (X).

ii. Make this direction parallel with vertical cross-wire, uncross the nicols (take out the analyser) and note the colour. This colour corresponds to the fast (X) vibration direction. Say the colour is blue.

iii. Rotate the stage by 90°. Now the slow (Z) vibration direction is parallel with vertical cross wire. Say the colour is red.

iv. Since in the Newton's scale, red comes first followed by blue, absorption for red (Z) is greater than blue (X). So the pleochroic (absorption) formula is Z > X or X < Z.

v. For biaxial minerals two sections, i.e. prismatic with one set of cleavage and pinacoidal or basal with two sets of cleavage are necessary to determine the scheme of pleochroism corresponding to X, Y and Z vibration directions. The prismatic section with one set of cleavage shows colours corresponding to fast (X) and slow (Z) vibration directions while the pinacoidal or basal section with two sets of cleavage shows colours corresponding to intermediate (Y) and fast/slow (X/Z) vibration directions.

For example, hypersthene shows pleochroism from pink (X) to green (Z). The pleochroic (absorption) formula is X < Z.

Some minerals have small spots which are more strongly pleochroic than the main body of the mineral. These are known as *pleochroic halos*. These are supposed to be produced by bombardment of radioactive emanations. These are characteristic of some minerals like biotite.

6.6.1.3 Refractive Index

Normally the minerals are permanently mounted in a Canada balsam medium. In isotropic minerals, which are single refracting, light travels with same velocity in all directions. These minerals have a single refractive index. The anisotropic minerals are doubly refracting. The component rays travel with different velocities in different directions. Uniaxial minerals have two refractive indices corresponding to ordinary and extraordinary rays, which are indicated by 'ω' and 'ε' respectively. Biaxial minerals, on the other hand, have three vibration directions fast (X), intermediate (Y) and slow (Z), corresponding to which there are three refractive indices designated as 'α', 'β' and 'γ'. In many instances the average refractive index of the mineral is determined by comparison with that of a set of liquids of known refractive indices. To determine the values of ω and ε in case of uniaxial minerals or α, β and γ in case of biaxial minerals specific oriented sections are necessary. However, these are not always available. Hence, the refractive indices of the minerals are generally determined relative to the mounting medium, i.e. Canada balsam of refractive index 1.537. To achieve this, the nature of outline, relief and surface of the mineral are studied. The distinctness of the outline depends upon the difference of the refractive indices of the mineral and the surrounding medium. If the mineral is of higher refractive index than the medium, the outline of the mineral appears distinct. In the opposite case, the outline appears faint. When the refractive index of the mineral is greater than that of glass or mounting medium, the rays seem to come from higher relief. This is due to the fact that a mineral of higher refractive index embedded in a mount of lower refractive index acts as a little lens. The light rays coming from the bottom surface appear to come from a slightly higher point. Such mineral seems to standout in relief compared to the surrounding medium. In case of opposite situation, i.e. if the mineral is of lower refractive index than the surrounding medium, the relief of the mineral appears low. The surface of a mineral section has minute elevations and depressions from which light is reflected at various angles. This results in pitted or rough nature of the mineral surface. The surface appears smooth if the mineral is of lower refractive index than the surrounding medium. The general rules are:

	Higher refractive index	Lower refractive index
Outline	Strong	Faint
Relief	High	Low
Surface	Rough	Smooth

In addition to the above, the relative refractive index of the mineral can be determined from the *Becke line*. When a bundle of rays are thrown onto the contact between two media, some of the rays are refracted and some are totally reflected resulting in a narrow band of light known as Becke line. When the objective is raised, the Becke line moves into the mineral of higher refractive index and when the objective is lowered, the Becke line moves into the mineral of lower refractive index. When the determination of exact refractive indices is not possible, the average refractive index of the mineral is estimated with respect to the mounting medium (Canada balsam). In such cases the refractive index is expressed as high or low with respect to Canada balsam.

6.6.1.4 Shape and Habit

Shape refers to the geometric form that a mineral grain takes by development in different dimensions. The shapes may be equant, bladed, lath shaped, etc as described in Section 5.1.6. The crystal habit is an expression of the shape like prismatic, tabular, etc. Euhedral, subhedral and anhedral are the terms used to describe crystal habit depending on the degree of development of faces. The term *euhedral* is applied to a grain showing well-developed faces or boundaries. Partial development of faces is known as *subhedral* while *anhedral* refers to complete absence of faces.

6.6.1.5 Cleavage

Cleavage appears as one or two sets of parallel lines in thin section (Fig. 6.5). The number of sets and their inclination with each other depends on the direction of section. For example, transverse section of hornblende shows two sets of cleavage inclined with each other by 120°, while longitudinal section of the same species displays only one set of cleavage. Some minerals like olivine and garnet generally do not show cleavage but are characterised by a network of cracks or fractures.

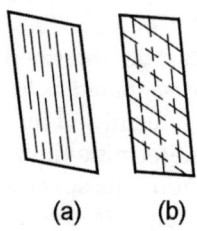

(a)   (b)

Fig. 6.5: Cleavage

6.6.1.6 Alteration

Alteration is the term applied to all changes in minerals subsequent to their formation. Normally alteration begins from cleavages or cracks and pervades deep into the mineral. An altered mineral commonly appears cloudy or turbid. Olivine is commonly altered to serpentine along cracks and some pyroxenes appear cloudy due to alteration.

6.6.1.7 Inclusion

Some minerals may enclose foreign substances like fluids or other crystals, which are known as inclusions. In some instances, as in case of apatite, these inclusions are regularly arranged in form of rows. Leucite contains volcanic glass; staurolite contains quartz grains; quartz contains bubbles, zircon crystals etc as inclusions.

6.6.1.8 Twinkling

The twinkling phenomenon is characteristic of calcite. The refractive indices of calcite for ordinary and extraordinary rays are 1.66 and 1.49 respectively while the refractive index of Canada balsam is 1.537. If a granular mosaic of calcite grains is examined in polarised light, some grains transmit ordinary ray while some grains transmit extraordinary ray. The grains transmitting ordinary ray have refractive index higher than the Canada balsam and, therefore, their boarders are well marked. On the other hand, grains transmitting extraordinary ray have refractive index less than the Canada balsam as a result of which their boarders are indistinctly marked. When the section is rotated, the grains show alternate strong and faint boarders resulting in the twinkling effect.

6.6.2 Properties under Crossed Nicols

In this case both the polariser and analyser are in use. Since the vibration directions of the light emerging out of the polariser and analyser are perpendicular to each other, the position is otherwise known as *crossed nicols*. In this situation, property of the mineral like optic character (isotropic or anisotropic, uniaxial or biaxial and positive or negative), interference colour, interference figure, birefringence, extinction, twinning, sign of elongation and optic axial angle are studied.

6.6.2.1 Optic Character (Isotropic or Anisotropic, Uniaxial or Biaxial and Positive or Negative)

Minerals crystallising in isometric system are single refracting and they have optical properties equal in all directions. The light rays leaving the polariser vibrate in left-right direction in case of synthetic polaroid or parallel to the shorter diagonal of the nicol prism. They pass through the isotropic mineral section without any change of the vibration direction and meet the analyser in top-bottom or parallel to the longer diagonal direction. Since the analyser allows the light vibrating in left-right or parallel to the shorter diagonal direction, the rays do not pass through the analyser and no light reaches the eye. As a result, the field of view becomes dark. If the mineral is crotated, the situation remains unchanged. Thus, under crossed nicols position, the isotropic minerals appear black in all positions.

The anisotropic minerals show interference colour in crossed nicols position. When rotated about 360°, the vibration directions of the mineral coincide four times with the vibration directions of the nicols. In these positions the light passes through the mineral without double refraction and the field of view becomes dark. The mineral is said to be in extinction position. So in a complete rotation of the stage, extinction takes place four times. Each anisotropic mineral has its own interference colour. For example, properly oriented quartz of 0.03 mm thickness shows first order yellow (straw yellow) interference colour. The interference colour shown by a mineral depends on the retardation of different wavelengths, which in turn depends on the orientation, birefringence and thickness of the mineral section. When these parameters change, the mineral shows variable interference colours. The interference colour changes its intensity and hue with rotation of the stage.

The anisotropic minerals are divided into two groups, uniaxial (minerals crystallising in teragonal and hexagonal systems) and biaxial (minerals crystallising in orthorhombic, monoclinic and triclinic systems). The uniaxial minerals have one direction (vertical, c-axis) along which both the ordinary and extraordinary rays travel with the same velocity, i.e. there is no double refraction. In case of biaxial minerals, there are two such directions. These two groups are distinguished from each other by their interference figures.

a. Obtaining an interference figure

 i. Focus the microscope on the mineral using high magnification in plane polarised light.
 ii. Insert the analyser to bring the mineral under crossed polarised light.
 iii. Fully open the sub-stage diaphragm and insert the condenser lens.
 iv. Insert the Betrand lens placed above the analyzer. In some cases a good interference figure is obtained by removing the eyepiece.
 v. In this position an interference figure is obtained, the nature of which depends on the mineral (uniaxial or biaxial) and direction of section with respect to optic axis.

b. Uniaxial interference figures

The uniaxial optic axis figure may be centered (Fig. 6.6) or off-centered (Fig. 6.7) or a flash-figure. The optic axis figure is obtained when the mineral section is perpendicular to the c-crystallographic axis (optic axis). The centered optic axis figure is the simplest of all. It consists of two intersecting black areas, known as isogyres and one or more sets of coloured rings, known as isochromes (Figs 6.6b and c).

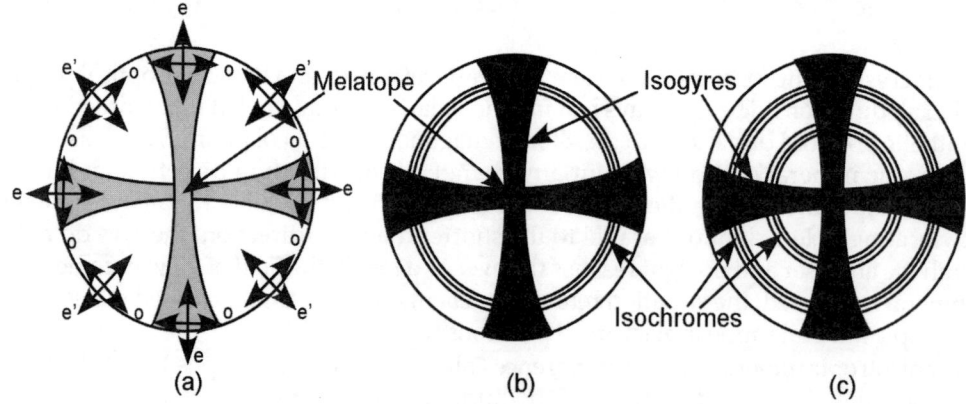

Fig. 6.6: Centered uniaxial interference figures

In Fig 6.6a, the arrows indicate the vibration directions of the ordinary (o) and extraordinary (e and e') rays. The extraordinary ray (E-ray) vibrates radially and the ordinary ray (O-ray) vibrates perpendicular to it. When these vibration directions are parallel with the vibration directions of the polariser and analyser, i.e. the cross-wires, no light passes and a dark cross (isogyres) result. The intersection point of the isogyres is known as *melatope*. The ray striking the mineral plate at normal incidence travels along the optic axis without undergoing double refraction. It cannot pass through the analyser. As a result, darkness results at melatope. The rays traveling oblique to optic axis are doubly refracted and the resulting rays have some path difference. They interfere producing interference colours. The path difference is zero along the optic axis and gradually increases outward as the rays become more and more inclined. Since the loci of the points where the path difference are equal are circular, concentric colour rings appear. The sets of colour rings (orders) depend on the birefringence of the mineral under examination. A mineral of low birefringence shows first-order colours (Fig. 6.6b), while minerals of high birefringence may show second- to fourth-order colours (Fig. 6.6c). If the optic axis of the crystal makes an angle with the axis of the microscope, the isogyres are asymmetrically located in the field of view (Fig. 6.7).

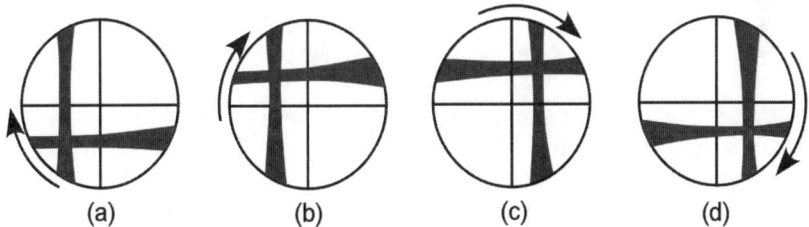

(a)   (b)   (c)   (d)

Fig. 6.7: Eccentric uniaxial interference figures

When the stage is rotated, the melatope moves in a circular path, but the isogyres remain parallel with the vibration directions of the polariser and analyser. If the inclination of the optic axis is so large that the melatope does not appear in the field of view, on rotation of the mineral, the isogyres move across the field maintaining their parallelism with the vibration directions of the polars. The 'flash figure' is an interference figure produced by a uniaxial mineral when the optic axis is normal to the axis of the microscope. When the mineral is at an extinction position the figure is an ill-defined cross, occupying much of the field. On rotation of the stage, the cross breaks into two hyperbolas, which rapidly leave the field in quadrants containing the optic axis.

c. Determination of optic sign of uniaxial minerals

In case of optic axis interference figure, the extraordinary ray (e-ray) vibrates radially (along the radius of the interference figure) and the ordinary ray (o-ray) vibrates tangentially to the isochromes (Fig. 6.6). Using the accessory plates, in which vibration directions of the rays are known, it is possible to know whether the e-ray is slower (i.e. mineral is positive) or faster (i.e. mineral is negative) than the o-ray and thus the optic sign of the mineral can be determined.

i. *By use of mica plate*: The ray vibrating along the length of mica plate is fast. When such a plate is superimposed on the optic axis interference figure of a positive mineral in which the e-ray vibrating parallel to slot (NW and SE quadrants) is slow, destructive interference takes place as a result of which the isochromes in the NW and SE quadrants shift towards the periphery and two black spots appear near the center. At the same time constructive interference takes place in the NE and SW quadrants resulting in increase of the order of interference colour, i.e. isochromes move towards the center (Fig. 6.8a). The situation becomes opposite in case of negative minerals and black spots appear in NE and SW quadrants (Fig. 6.8b).

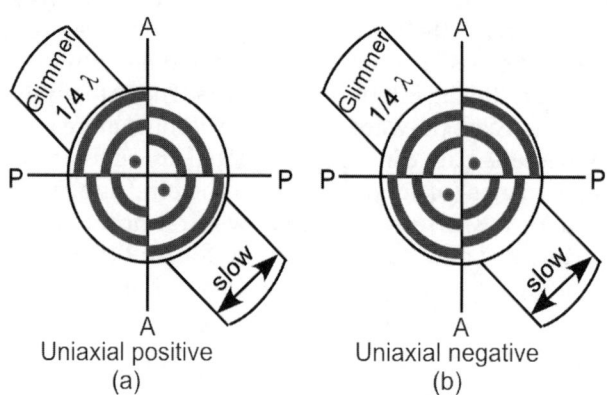

Uniaxial positive   Uniaxial negative
(a)   (b)

Fig. 6.8: Determination of optic sign by mica plate

ii. By use of gypsum plate: The gypsum plate is normally used to determine the optic sign when low-order interference colours or no colours are seen in the optic axis figure. The ray vibrating along the length of gypsum plate is fast. This plate has the effect of superimposing red of first order on the interference figure. In the quadrants where there is addition (constructive interference), red plus gray gives blue; in the alternate quadrants, destructive interference takes place where red minus gray gives yellow. The arrangement of colours in positive mineral is: Yellow NW-SE; blue NE-SW (Fig. 6.9a) and in negative mineral is: Yellow NE-SW, blue NW-SE (Fig. 6.9b). Since, in case of positive minerals, slow e-ray vibrating parallel to slot (NW and SE quadrants) is superposed by fast ray of gypsum plate, destructive interference takes place giving rise to yellow colour in NW and SE quadrants (Fig. 6.9a). In case of negative minerals the situation is opposite.

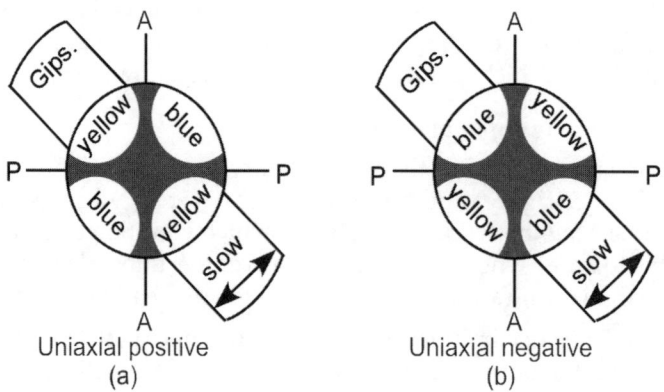

Uniaxial positive   Uniaxial negative
(a)   (b)

Fig. 6.9: Determination of optic sign by gypsum plate

iii. By use of quartz wedge: The quartz wedge is most effective in determining optic sign when high-order interference colours are seen in optic axis figure. The ray vibrating along the length of wedge is fast. The wedge is usually inserted with the thin edge forward. As the wedge moves through the microscope tube, constructive interference takes place in two opposite quadrants and destructive interference takes place in remaining quadrants progressively. As a result, colours move inward/outward in different quadrants. In case of positive mineral, colour rings move away from the center in NW-SE quadrants and towards the center in NE-SW quadrants (Fig. 6.10a). In case of negative mineral, colour rings move towards the center in NW-SE quadrants and away from the center in NE-SW quadrants (Fig. 6.10b). However, the situation becomes opposite if the ray vibrating along the length of the accessory plate is slow.

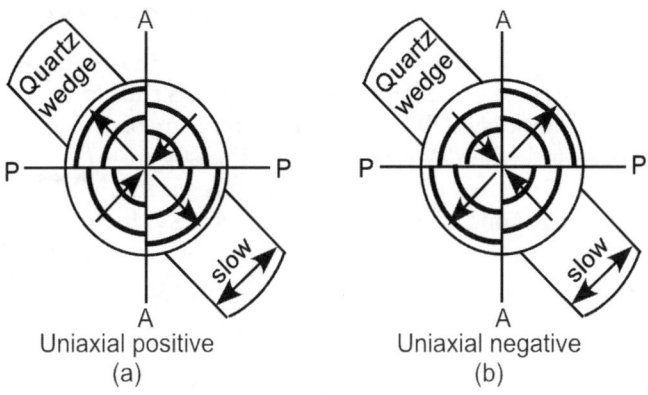

Uniaxial positive
(a)

Uniaxial negative
(b)

Fig. 6.10: Determination of optic sign of uniaxial mineral by quartz wedge

d. Biaxial interference figures

The biaxial minerals have three vibration directions along which light travels with different velocities. The vibrations of the fastest ray (X) and slowest ray (Z) are perpendicular to each other. The direction perpendicular to the X–Z plane (optic plane) is the vibration direction of the intermediate ray (Y) (Fig. 6.11a, b). This direction coincides with the optic normal (ON). The biaxial indicatrix (Fig. 6.11c) is a triaxial ellipsoid with its three axes mutually perpendicular directions X, Y and Z. There are two circular sections (S, Fig. 6.11c) perpendicular to which the velocities of the fast and slow rays are equal. These perpendicular directions are known as the optic axes (OA). The acute angle between optic axes is known as 2V. In positive biaxial mineral the slow ray (Z) is the acute bisectrix (Bxa) and the fast ray (X) is the obtuse bisectrix (Bxo) (Fig. 6.11a). In negative biaxial mineral the fast ray (X) is the acute bisectrix (Bxa) and the slow ray (Z) is the obtuse bisectrix (Bxo) (Fig. 6.11b).

Depending on the direction of mineral section four types of biaxial interference figures are obtained.

i. Optic axis figure: The optic axis figure (Fig. 6.12) is observed in minerals cut perpendicular to an optic axis. The figure consists of a single isogyre at the center of which the optic axis is located. When the optic plane (XZ plane) is parallel to the vibration direction of any polar, the isogyre crosses the center of the field as a straight bar. On rotation of the stage it swings

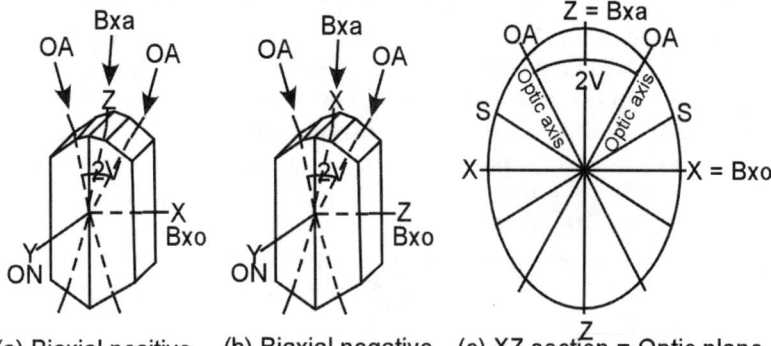

(a) Biaxial positive   (b) Biaxial negative   (c) XZ section = Optic plane

Fig. 6.11: Vibration directions and optic axes in biaxial minerals

across the field forming a hyperbola at 45° position. At this position the convex side of the isogyre points towards the acute bisectrix. As the amount of 2V increases the curvature of the isogyre decreases and when 2V is equal to 90°, the isogyre is straight. Figure 6.12 shows a number of isogyres with different curvature corresponding to different values of 2V. It can be used as a reference figure to calculate the value of 2V.

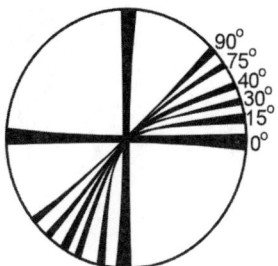

Fig. 6.12: Optic axis figure of biaxial mineral showing curvature of the isogyres for different values of 2V

ii. *Acute bisectrix figure*: The acute bisectrix figure (Fig. 6.13) is obtained in a mineral cut perpendicular to acute bisectrix (Bxa, Fig. 6.11). If the 2V is very small, there are four positions in a complete rotation of the stage at which the figure resembles the uniaxial optic axis figure. In these positions, the figure consists of isochrome lines and isogyres perpendicular to each other forming a black cross (Fig. 6.13a). When the stage is rotated, the black cross breaks into two hyperbolas with maximum separation at 45° position and the isochromatic curves assume oval shape (Fig. 6.13b). When 2V exceeds 60°, the isogyres leave the field; larger the optic angle, faster they leave.

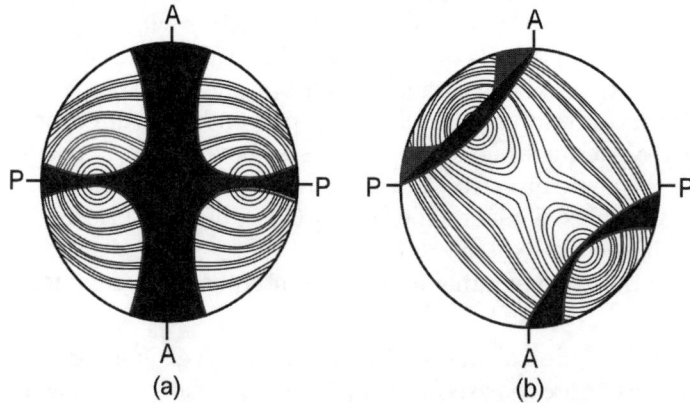

(a)   (b)

Fig. 6.13: Two views of the acute bisectrix (Bxa) figure: (a) Optic plane is parallel to polariser and analyser, (b) Optic plane is inclined to polariser and analyser

iii. Obtuse bisectrix figure: The obtuse bisectrix figure is obtained in a mineral cut perpendicular to obtuse bisectrix (Bxo, Fig. 6.11). When the plane of the optic axis is parallel to the vibration direction of any polar, the isogyres form a black cross. On rotation of the stage, the cross breaks into two isogyres that move rapidly out of the field in the direction of the acute bisectrix. The figure is similar to one shown in Fig. 6.13.

iv. Optic normal figure: The optic normal figure is obtained in sections cut parallel to the plane of the optic axis (perpendicular to optic normal, ON, Fig. 6.11). When the X and Z directions are parallel to the vibration directions of the polars, the figure is a poorly defined cross. On slight rotation of the stage, the cross splits into hyperbolas that rapidly leave the field in the quadrants containing the acute bisectrix. The biaxial optic normal figures appear similar to uniaxial flash figures.

e. Determination of optic sign of biaxial minerals

The sign of the biaxial mineral can be best determined from acute bisectrix (Bxa) or optic axis (OA) figures. In case of Bxa figure, turn the stage to 45° position so that the isogyres are in the NW-SW quadrants and insert the gypsum plate in which fast ray vibrates parallel to length. In positive minerals the slow ray (Z) is the acute bisectrix. Due to destructive and constructive interferences

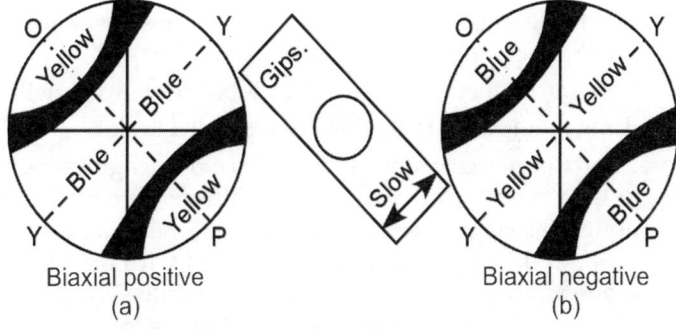

Biaxial positive
(a)

Biaxial negative
(b)

Fig. 6.14: Determination of optic sign from acute bisectrix figure

yellow colour is produced on the concave side and blue colour is produced on the convex side of the isogyres (Fig. 6.14a). In case of negative minerals the fast ray (X) is the acute bisectrix. With the gypsum plate, blue colour is produced on the concave side and yellow colour is produced on the convex side of the isogyres (Fig. 6.14b). In a similar manner, the optic axis (OA) figure in the 45° position can be used to determine the optic sign. With gypsum plate, the positive mineral shows blue on convex side and yellow on concave side (Fig. 6.15a) while the negative mineral shows yellow on convex side and blue on concave side (Fig. 6.15b).

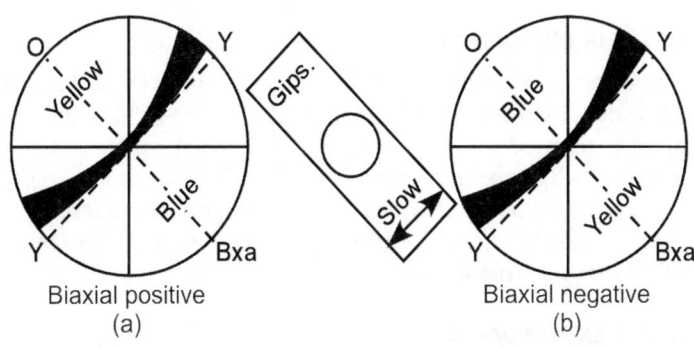

Biaxial positive
(a)

Biaxial negative
(b)

Fig. 6.15: Determination of optic sign from optic axis figure

The quartz wedge can be used to determine the optic sign if several isochrome lines are seen. As the wedge is inserted, the isochromes move in where gypsum plate shows blue colour and move out where gypsum plate shows yellow colour.

6.6.2.2 Interference Colour

Since the isotropic minerals are single refracting they do not show any interference colour and extinct in all positions under crossed nicols. The colours shown by the anisotropic (doubly refracting) minerals between cross polars are known as their interference colours. The method of the production of interference colours has been described in Section 6.3.3 and the colours are listed in Tables 6.1 and 6.2. Certain minerals like biotite, chlorite, serpentine, zoicite show abnormal or anomalous interference colours not found in Newton's chart. Colours like gray and white belong to first order, while colours like blue, green, yellow, red, etc. repeat several times with increase of path difference. Thus it is pertinent to classify them into different orders (Table 6.2). The order of the interference colour can be conveniently determined by quartz wedge. The procedure is outlined below:

i. Bring the mineral to extinction position. At this position the vibration directions of the mineral are parallel with those of polars.

ii. Rotate the stage anticlockwise by 45°. At this position one of the vibration directions of the mineral is parallel to slot (NW-SE).

iii. Determine whether this direction is fast or slow by methods indicated in Section 6.5.

iv. Bring the slow vibration direction parallel to slot.

v. Insert the quartz wedge (which is fast along length) slowly and observe the change of colour of the mineral. Due to destructive interference the colour will fall to first-order gray. By noting the successive change of colour, the order can be determined. For example, if the original interference colour is red and with insertion of the quartz wedge the colour sequentially changes as: Red – yellow – white – gray, the original interference colour is first-order red. If the list reads: Red – yellow – green – blue – red – yellow – white – gray, the original interference colour is second-order red.

The alternative method is to take out the mineral section when the colour has changed to first-order gray. The quartz wedge will show the original interference colour shown by the mineral. If the eyepiece is also taken out, colour bands will appear. Count the number of red bands (n) while taking out the quartz wedge slowly. The order of the interference colour is (n + 1).

6.6.2.3 Birefringence

Birefringence is a measure of the difference between the highest and lowest refractive indices of a mineral. In case of uniaxial minerals it is '$\varepsilon \sim \omega$' and in case of biaxial minerals it is '$\alpha \sim \gamma$'. It can be estimated by measuring the individual values of the refractive indices. Alternatively, since this is a factor that controls the interference colour of the mineral, it can be conveniently read from the *Michel-Lavy chart*. The approximate values of the birefringence corresponding to different interference colours are given in Table 6.3. These values are for perfectly cut sections of 0.03 mm thickness.

6.6.2.4 Extinction

The vibration directions of the minerals coincide with those of polars four times in a complete rotation of the stage (360°) as a result of which anisotropic minerals extinct four times in a complete rotation of the stage under crossed nicol. In these positions, the polarised light coming out of the polariser passes through the mineral section without double refraction, meets the analyser in a direction perpendicular to the vibration direction of light in the analyser and is cut

Table 6.3: Scale of birefringence

Order	Colour	Birefringence	Order	Colour	Birefringence
First	Gray	0.004	Third	Blue	0.033
	White	0.007		Green	0.036
	Yellow	0.009		Yellow	0.039
	Orange	0.012		Orange	0.042
	Red	0.015		Red	0.045
	Violet	0.017		Violet	0.048
Second	Blue	0.019	Fourth	Blue	0.051
	Green	0.022		Green	0.054
	Yellow	0.025		Yellow	0.057
	Orange	0.027		Pink	0.060
	Red	0.029			
	Violet	0.030			

off without reaching the eye. In these positions, the field of view become dark and the mineral is said to be in extinction position. The extinction angle is the acute angle between the vibration direction (X or Z) and crystallographic direction like cleavage, crystal face, twin plane or mineral edge. The extinction angle is one of the important parameters used for identification of the mineral. The method of determination of the extinction angle is given below. Before attempting to measure the extinction angle it should be ensured that the microscope is free from eccentricity and the mineral possesses some crystallographic direction.

i. Under crossed nicol bring the mineral to extinction position. In many cases the mineral passes through a dark gray stage before becoming completely black. The human eye is not sensible to discriminate between no light and very less light. To ascertain the exact extinction position the gypsum plate is inserted into the slot. At the exact extinction position, the gypsum plate gives a sensible violet colour. With slight change in the extinction position, the quality of the violet colour changes, which can be detected by the eye. At this position one of the vibration directions of mineral is parallel to vertical cross-wire. Note the stage reading (R_1).

ii. Rotate the stage by 45° in anticlockwise direction. Now the vibration direction that was parallel with the vertical (N–S) cross-wire is parallel to the slot. By inserting the mica plate (which is fast along the length) determine the fast (X) or slow (Z) nature of the vibration direction of the mineral parallel to slot. (See Section 6.5)

iii. Take out the analyser and bring a crystallographic direction (e.g. cleavage) parallel to vertical cross-wire. Note the stage reading (R_2).

iv. The difference between the stage readings ($R_1 \sim R_2$) is the extinction angle ($X \wedge C$ or $Z \wedge C$). If $X \wedge C$ is less than 45°, $Z \wedge C$ will be more than 45° and vice versa. The least value is commonly taken as the extinction angle.

The extinction is also expressed qualitatively by terms like *straight* or *parallel*, *inclined* or *oblique* and *symmetrical*. In case of straight or parallel extinction, extinction takes place when the crystallographic direction, cleavage, etc. is parallel with a cross-wire. In this case the amount of extinction angle is 0°. Inclined or oblique extinction occurs when the crystallographic direction, cleavage, etc. make an angle with the cross-wires in the extinction position. In this case the amount of extinction angle varies from 1° to 45°. In symmetrical extinction, the cross-wires bisect the angle between two sets of cleavages or edges in the extinction position. It may be noted that the vibration directions of the polars are coincident with the cross-wires.

6.6.2.5 Sign of Elongation

Many minerals are habitually elongated in a particular direction in preference to others. Many of the uniaxial minerals are elongated in the direction of vertical crystallographic axis. Determination of sign of elongation is relatively straightforward. If the slow ray is parallel or nearly so with the direction of elongation, the mineral is said to be *length slow* and the sign of elongation is *positive*. In the contrary, if the fast ray is parallel or nearly so with the direction of elongation, the mineral is said to be *length fast* and the sign of elongation is *negative*. It is only necessary to determine the slow (Z) and fast (X) vibration directions of the mineral by the method outlined in Section 6.5 and to examine which one of them is parallel or more parallel with the direction of elongation.

6.6.2.6 Twinning

A twin crystal consists of two or more individual crystals, which show parallelism in certain parts and at the same time other parts are differently oriented. Twinning is best observed under crossed nicol position and in some cases it forms important criteria for mineral identification. Since the individual units are differently oriented, they exhibit different interference colours and go to extinction position in different orientations. Figure 6.16a shows plagioclase feldspar in plane polarised light. Under crossed nicols, it shows lamellar twinning, as the individual parts extinct

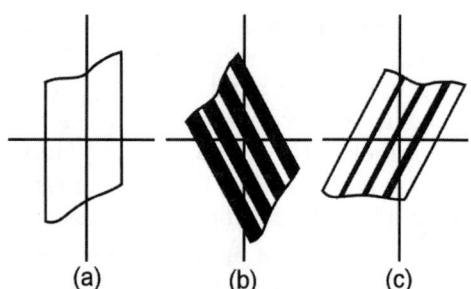

(a)   (b)   (c)

Fig. 6.16: Lamellar twinning in plagioclase feldspar

in different orientations (Figs 6.16b and c). Microcline and orthoclase show cross-hatchet (net like) and simple twinning respectively.

6.6.2.7 Optic Axial Angle (2V)

Biaxial minerals have two optic axes along which both ordinary and extraordinary rays travel with same velocity, i.e. there is no double refraction. The acute angle between these axes is known as optic axial angle or 2V. It can be estimated from the optic axial figure (Fig. 6.12), refractive indices of the mineral by the formula

$$\tan^2 V = \frac{\dfrac{1}{\alpha^2} - \dfrac{1}{\beta^2}}{\dfrac{1}{\beta^2} - \dfrac{1}{\gamma^2}}$$ or by actual measurement by universal stage.

6.7 MICROSCOPIC IDENTIFICATION OF ROCK-FORMING MINERALS

Since most of the rock-forming minerals are transparent, refracting type of petrological microscope is necessary for their examination. The mineral slide to be studied is kept on the rotating stage of the microscope and firmly fixed by stage clips.

6.7.1 Examination in Plane-polarised Light

Properties like colour and pleochroism, refractive index, cleavage, shape and habit, alteration, inclusion and twinkling are studied in plane-polarised light.

6.7.1.1 Colour and Pleochroism

The colour is a distinctive character of a mineral. The mineral may be colourless or coloured that does not change with rotation of the stage. In this case, the mineral is said to be nonpleochroic. It may show change of colour when rotated with the stage owing to its pleochroic nature. In this case, the pleochroic nature and colors of the mineral are noted. For example, actinolite is pleochroic from yellow to green. In case of pleochroic minerals, both the colours are to be noted.

6.7.1.2 Refractive Index

Isotropic minerals are single refracting; uniaxial minerals have two indices of refraction and biaxial minerals have three indices of refraction. For later two cases, specially oriented sections are necessary, which are not always available. The average refractive index of a mineral can be quantitatively determined by Becke test, if refractive index liquids are available. In many instances average refractive index is determined qualitatively with respect to the mounting medium, i.e. Canada balsam of refractive index 1.537 by studying the nature of outline, relief and surface of the mineral in addition to the movement of Becke line by the criteria mentioned in Section 6.6.1.3. For example, the refractive indices of gypsum and zircon are lower and higher than Canada balsam (1.537) respectively.

6.7.1.3 Shape and Habit

Shape and habit of a mineral can be ascertained if the mineral section is entire, i.e. it represents the total mineral grain. After viewing the entire section in low magnification, a suitable term as mentioned in Section 6.6.1.4 and Table 5.5 is assigned. Further, any one of the terms like euhedral, subhedral or anhedral can be used depending on the degree of growth of crystal faces. For example, andalusite shows columnar shape and euhedral habit.

6.7.1.4 Cleavage

In thin section, cleavage appears as one or more sets of continuous or discontinuous parallel lines. The number of cleavage lines and their inclination with each other depends on the direction of section. For example, a longitudinal section of hornblende shows one set of cleavage while the prismatic section of the same mineral shows two sets of cleavages inclined at 120°. Mineral like quartz is devoid of cleavage, whereas minerals like olivine and garnet show a number of irregular cracks instead of cleavage.

6.7.1.5 Alteration

An altered mineral looks turbid or cloudy and generally alteration proceeds from cleavages and cracks in the mineral. Since in the process of alteration, a homogeneous mineral is converted into a large number of irregularly arranged grains, they are better visualized by differences of interference colour under crossed nicols. For example, different varieties of feldspars alter to clay minerals, whereas biotite and diopside alter to chlorite.

6.7.1.6 Inclusion

Inclusions like fluids and other minerals are readily identified by their difference of optical properties from the enclosing minerals. In some minerals the inclusions show a definite pattern. For example, in apatite inclusions are arranged in rows and in leucite inclusions occur in concentric or radial patterns. In many cases presence of inclusions and their patterns are used as diagnostic property for mineral identification.

6.7.1.7 Twinkling

Very few minerals show the twinkling effect. To show this property, the anisotropic mineral should consist of an aggregate of crystal units with one refractive index less and another refractive index more than that of the mounting medium. Calcite is the best known example.

6.7.2 Examination Under Crossed Nicols

In this case the analyser is in use and the accessory plates are inserted into the slot whenever necessary. Optic characters like isotropism or anisotropism, uniaxial or biaxial, positive or negative, interference colour, birefringence, extinction angle, sign of elongation, twinning and optic axial angle are studied under crossed nicols.

6.7.2.1 Optic Character

If a mineral section appears black in all positions of the stage, it is said to be isotropic. Minerals crystallising in isometric system belong to this category. However, care should be taken in study of basal sections of uniaxial minerals, which behave in similar manner. Detrital grains of biotite in heavy mineral study are often isotropic.

If the mineral shows some colour and becomes black four times in a complete rotation of the stage, the mineral is anisotropic. An anisotropic mineral may be either uniaxial or biaxial, which is ascertained from the nature of interference figure. Minerals crystallising in tetragonal and hexagonal systems exhibit uniaxial interference figures. Basal section cut perpendicular to vertical crystallographic axis (c) gives the most suitable interference figure that consists of a black cross made by the intersection of two isogyres and a series of coloured rings. If the optic axis of the crystal makes an angle with the axis of the microscope, i.e. the section is inclined to the vertical crystallographic axis, the isogyres are not symmetrically located in the field of view. With rotation of the stage, they move quadrant-wise with the melatope. Minerals crystallising in orthorhombic, monoclinic and triclinic systems are biaxial. The section cut perpendicular to the acute bisectrix of biaxial mineral shows two intersecting isogyres and colour rings arranged about two eyes when the optic plane is parallel to one of the polars. With rotation of the stage, the isogyres separate into two hyperbolas one passing through each eye.

The positive and negative character of the anisotropic mineral depends on the relative velocities of the ordinary and extraordinary rays within the mineral. This can be ascertained by superposing the accessory plates on mineral section and recording the changes of interference colours as mentioned in Section 6.6.2.1.

6.7.2.2 Interference Colour

Due to the interference of two extraordinary rays leaving the analyser, the section of an anisotropic mineral appears coloured when viewed under crossed nicols. The basal section of uniaxial mineral is an exception. The intensity of the colour is maximum in the mid-way between extinction positions. If the colour of the mineral is other than gray and white, it is necessary to determine the order of the interference colour. It is achieved by compensating the interference colour of the mineral by a quartz wedge (Section 6.6.2.2).

6.7.2.3 Birefringence

The birefringence is the difference between the highest and lowest refractive indices of a mineral. If required numbers of oriented sections are available, the refractive indices of the mineral with respect to ordinary and extraordinary rays can be accurately determined and the exact value of birefringence can be found out. This is, however, not always possible. The interference colour shown by the mineral is directly related to birefringence provided the section is exactly 0.03 mm thick. The approximate value of the birefringence can be determined from the interference colour (Table 6.3).

6.7.2.4 Extinction

The extinction of a mineral may be straight or parallel (extinction angle $0°$), oblique (extinction angle $1°-45°$) or symmetrical (optical direction bisects the angle between two sets of cleavages). The amount of the extinction angle and the position of the optic direction (X or Z) can be conveniently determined by the procedures mentioned in Section 6.6.2.4.

6.7.2.5 Sign of Elongation

Some minerals are habitually elongated in certain direction in preference to others. The first and slow vibration directions of the mineral can be determined by the procedures mentioned in Section 6.5. If the slow vibration direction is parallel or nearly parallel to the direction of elongation, the sign of elongation of the mineral is said to be positive. In the opposite case, the sign of elongation of the mineral is negative. For example, the sign of elongation of actinolite is positive while that of aegirine is negative.

6.7.2.6 Twinning

Twinning is best studied in crossed nicol position. In a twinned mineral, the individual units are differently oriented as a result of which they exhibit different interference colours in the same position of the mineral section and individual units go to extinction position in different orientations of the mineral section. Best known examples are plagioclase feldspars, which show lamellar twinning (Fig. 6.16) and microcline that shows cross-hatchet twinning.

6.7.2.7 Optic Axial Angle (2V)

Determination of optic axial angle requires a universal stage and oriented mineral sections. If the three refractive indices α, β and γ of the biaxial mineral can be determined accurately from differently oriented sections, the amount of optic axial angle can be calculated by the formula:

$$\tan^2 V = \frac{\dfrac{1}{\alpha^2} - \dfrac{1}{\beta^2}}{\dfrac{1}{\beta^2} - \dfrac{1}{\gamma^2}}$$

A rough estimation of the optic axial angle can be made from the optic axial figure (Fig. 6.12) The distinguishing characters of different rock forming minerals are given in Table 6.4.

Table 6.4: Optical characters of rock forming minerals

	Mineral	Actinolite	Aegirine	Albite	Allanite
Plane polarized light	Pleochroism/colour	Pleochroic, yellow to green	Pleochroic, yellow to green	Nonpleochroic, colourless	Pleochroic, brown
	Refractive index	1.61–1.66 (high)	1.745–1.836 (high)	1.527–1.542	1.64–1.80 (high)
	Shape and habit	Prismatic, fibrous	Prismatic, bladed	Plates, lath-shaped	Granular, columnar
	Cleavage	2 sets perfect (56°)	2 sets perfect (87°)	1 set perfect	Imperfect
	Alteration	Talc	Uncommon	Clay minerals	Amorphous substance
	Inclusion	Uncommon	Uncommon	Uncommon	Uncommon
Crossed nicol	Optic character	Biaxial negative	Biaxial negative	Biaxial positive	Biaxial negative
	Interference colour	Second order yellow	Third-fourth order	First order gray	Masked by body colour
	Birefringence	0.022–0.027	0.037–0.059	0.009–0.011	0.010–0.030
	Extinction	10°–20°	2°–10°	12°–19°	Parallel
	Twinning	Polysynthetic	Absent	Polysynthetic, lamellar	Absent
	Sign of elongation	Positive	Negative	–	–
	Optic axial angle	70°–85°	60°–66°	77°–82°	–

	Mineral	Alunite	Analcite	Anatase	Andalusite
Plane polarized light	Pleochroism/colour	Nonpleochroic, colourless	Nonpleochroic, colourless	Weakly pleochroic, yellow, brown, blue	Pleochroic, rose red to pale green
	Refractive index	1.57–1.59 (high)	1.487 (low)	2.48–2.56 (high)	1.629–1.647 (high)
	Shape and habit	Tabular aggregates	Equant	Tabular, striated	Euhedral, columnar
	Cleavage	1 set perfect	Imperfect	Imperfect	2 sets perfect
	Alteration	Uncommon	Uncommon	Uncommon	Sillimanite or sericite
	Inclusion	Uncommon	Uncommon	Uncommon	Carbonaceous matter
Crossed nicol	Optic character	Uniaxial positive	Isotropic	Uniaxial negative	Biaxial negative
	Interference colour	Second order blue	–	Third order colours	First order yellow
	Birefringence	0.020	–	0.073	0.007–0.011
	Extinction	Parallel	–	Parallel	Parallel
	Twinning	Absent	–	Absent	Absent
	Sign of elongation	–	–	–	Negative
	Optic axial angle	–	–	–	75°–85°

(Contd....)

	Mineral	Andesine	Anhydrite	Anorthite	Anorthoclase
Plane polarized light	Pleochroism/colour	Nonpleochroic, colourless	Nonpleochroic, colourless	Nonpleochroic, colourless	Nonpleochroic, colourless
	Refractive index	1.543–1.562 (high)	1.57–1.61 (high)	1.573–1.590 (high)	1.522–1.541
	Shape and habit	Euhedral to anhedral	Subhedral	Lath shaped	Anhedral
	Cleavage	1 set perfect	2 sets perfect	1 set perfect	1 set perfect
	Alteration	Clay minerals	Gypsum	Uncommon	Clay minerals
	Inclusion	Uncommon	Uncommon	Uncommon	Albite, sanidine
Crossed nicol	Optic character	Biaxial +/–	Biaxial positive	Biaxial negative	Biaxial negative
	Interference colour	First order white	Third order green	First order gray	First order gray
	Birefringence	0.008	0.044	0.012	0.005–0.007
	Extinction	13°–27°	Parallel	51°–70°	1°–10°
	Twinning	Polysynthetic, lamellar	Polysynthetic	Polysynthetic, lamellar	Polysynthetic
	Sign of elongation	–	–	–	–
	Optic axial angle	76°–90°	44°	77°–79°	43°–54°

	Mineral	Antigorite	Anthophyllite	Apatite	Aragonite
Plane polarized light	Pleochroism/colour	Colourless to pale green	Nonpleochroic, colourless	Nonpleochroic, colourless	Nonpleochroic, colourless
	Refractive index	1.555–1.573 (high)	1.598–1.676 (high)	1.630–1.655 (high)	1.530–1.686 (high)
	Shape and habit	Fibro-lamellar	Columnar, fibrous	Euhedral, prismatic	Columnar, fibrous
	Cleavage	Imperfect	2 sets	Imperfect	Imperfect
	Alteration	Uncommon	Talc	Uncommon	Calcite
	Inclusion	Uncommon	Absent	Inclusions in rows	Uncommon
Crossed nicol	Optic character	Biaxial negative	Biaxial positive	Uniaxial negative	Biaxial negative
	Interference colour	First order yellow	Second order	First order gray	High order pearl gray
	Birefringence	0.007–0.009	0.030	0.003–0.004	0.155
	Extinction	Parallel	Parallel	Parallel	Parallel
	Twinning	Absent	Absent	Absent	Lamellar
	Sign of elongation	Positive	Positive	Negative	–
	Optic axial angle	20°–60°	65°–90°	–	18°

(Contd...)

	Mineral	Augite	Axinite	Barite	Beryl
Plane polarized light	Pleochroism/colour	Nonpleochroic, colourless	Feebly pleochroic, colourless to pale violet	Nonpleochroic, colourless	Nonpleochroic, colourless
	Refractive index	1.686–1.737 (high)	1.678–1.696 (high)	1.63–1.64 (high)	1.564–1.598 (high)
	Shape and habit	Subhedral, prismatic	Anhedral	Elongate	Prismatic
	Cleavage	2 sets perfect	Imperfect	2 sets perfect	Imperfect
	Alteration	Hornblende, epidote	Uncommon	Uncommon	Kaolin
	Inclusion	Uncommon	Common	Uncommon	Liquid
Crossed nicol	Optic character	Biaxial positive	Biaxial negative	Biaxial positive	Uniaxial negative
	Interference colour	Second order	First order red	First order yellow	First order yellow
	Birefringence	0.021–0.025	0.010–0.012	0.012	0.004–0.008
	Extinction	36°–45°	Oblique	Parallel	Parallel
	Twinning	Polysynthetic	Absent	Polysynthetic	Absent
	Sign of elongation	–	–	–	Positive
	Optic axial angle	58°–67°	70°–75°	36°–37°	–

	Mineral	Biotite	Bohemite	Brookite	Brucite
Plane polarized light	Pleochroism/colour	Pleochroic, brown	Nonpleochroic, colourless	Weakly pleochroic, orange, brown	Nonpleochroic, colourless
	Refractive index	1.57–1.61 (high)	1.638–1.651 (high)	2.58–2.72 (high)	1.566–1.585 (high)
	Shape and habit	Tabular	Tabular	Euhedral, striated	Platy, scaly
	Cleavage	1 set perfect	1 set	Striations	1 set perfect
	Alteration	Chlorite	Uncommon	Uncommon	Hydromagnesite
	Inclusion	Zircon, pleochroic halos	Uncommon	Uncommon	Uncommon
Crossed nicol	Optic character	Biaxial negative	Biaxial negative	Biaxial positive	Biaxial positive
	Interference colour	Second order (often masked by body colour)	Second order colours	High order brown	First order red
	Birefringence	0.033–0.059	0.013	0.14	0.020
	Extinction	Parallel	Parallel	Incomplete	Parallel
	Twinning	Mica law	Absent	Absent	Absent
	Sign of elongation	Positive	Positive	–	Negative
	Optic axial angle	0°–25°	moderate	0°–30°	–

(Contd...)

	Mineral	Calcite	Cancrinite	Celestite	Chabazite
Plane polarized light	Pleochroism/colour	Nonpleochroic, colourless, cloudy	Nonpleochroic, colourless to pale yellow	Nonpleochroic, colourless	Nonpleochroic, colourless
	Refractive index	1.486–1.658	1.496–1.524 (low)	1.622–1.631 (high)	1.478–1.490 (low)
	Shape and habit	Rhombohedral, oolitic, spherulitic	Anhedral	Euhedral, tabular	Euhedral
	Cleavage	2 sets perfect	1 set	1 set perfect	Imperfect
	Alteration	Replaced by quartz	Present	Uncommon	Uncommon
	Inclusion	Present	Uncommon	Uncommon	Uncommon
Crossed nicol	Optic character	Uniaxial negative	Uniaxial negative	Biaxial positive	Biaxial positive
	Interference colour	Higher order, twinkling	First and second order colours	First order yellow	First order gray
	Birefringence	0.172	0.007–0.028	0.009	0.002–0.010
	Extinction	Symmetrical	Parallel	Parallel	Symmetrical
	Twinning	Polysynthetic	Absent	Absent	Absent
	Sign of elongation	–	Negative	Positive	–
	Optic axial angle	–	–	51°	0°–32°

	Mineral	Chlorite	Chloritoid	Chondrodite	Chrysotile
Plane polarized light	Pleochroism/colour	Weakly pleochroic, green to yellowish green	Pleochroic, shades of green and gray	Pleochroic, pale brown to redish brown	Nonpleochroic, colourless to green
	Refractive index	1.56–1.61 (high)	1.715–1.737 (high)	1.592–1.670 (high)	1.493–1.557
	Shape and habit	Subhedral, tabular	Subhedral, tabular	Subhedral	Fibrous
	Cleavage	1 set	1 set perfect	Parting present	Imperfect
	Alteration	Clay minerals	Clay minerals	Uncommon	Uncommon
	Inclusion	Uncommon	Present (hourglass)	Uncommon	Uncommon
Crossed nicol	Optic character	Biaxial negative	Biaxial positive	Biaxial positive	Biaxial positive
	Interference colour	Abnormal colours (royal blue)	First order colours	Second order green, red	First order yellow
	Birefringence	0.006–0.020	0.013–0.016	0.027–0.035	0.011–0.014
	Extinction	Parallel	0°–20°	26°–31°	Parallel
	Twinning	Absent	Polysynthetic	Polysynthetic	Absent
	Sign of elongation	–	Negative	Negative	Positive
	Optic axial angle	0°–40°	45°–65°	60°–90°	0°–50°

(Contd...)

Mineral		Clinozoisite	Cordierite	Corundum	Cristobalite	Cummingtonite
Plane polarized light	Pleochroism/colour	Pleochroic, colourless to pink	Nonpleochroic, colourless	Nonpleochroic, colourless	Nonpleochroic, colourless	Feebly pleochroic, colourless
	Refractive index	1.696–1.720 (high)	1.532–1.570 (high)	1.759–1.772 (high)	1.484–1.487 (low)	1.639–1.686 (high)
	Shape and habit	Euhedral, columnar	Anhedral, prismatic	Euhedral, tabular, prismatic	Squarish	Prismatic
	Cleavage	1 set perfect	Imperfect	Parting present	Curved fracture	2 sets perfect
	Alteration	Uncommon	Sericite, chlorite, talc	Uncommon	Uncommon	Uncommon
	Inclusion	Uncommon	Common with pleochroic halos	Uncommon	Uncommon	Uncommon
Crossed nicol	Optic character	Biaxial positive	Biaxial +/−	Uniaxial negative	Uniaxial negative	Biaxial positive
	Interference colour	First order colours	First order yellow	Second order	First order gray	Second order
	Birefringence	0.005–0.015	0.007–0.011	0.008	0.003	0.025–0.029
	Extinction	Parallel	Parallel	Parallel	Undulatory	15°–20°
	Twinning	Polysynthetic	Interpenetration	Lamellar	Absent	Polysynthetic
	Sign of elongation	Positive	–	Positive	Positive	Positive
	Optic axial angle	30°–60°	40°–80°	–	–	68°–87°

Mineral		Diaspore	Diopside	Dolomite	Enstatite
Plane polarized light	Pleochroism/colour	Nonpleochroic, colourless to pale blue	Nonpleochroic, colourless	Nonpleochroic, colourless	Nonpleochroic, colourless
	Refractive index	1.702–1.750 (high)	1.65–1.727 (high)	1.50–1.67 (high)	1.650–1.674 (high)
	Shape and habit	Tabular	Subhedral, prismatic	Subhedral	Prismatic
	Cleavage	1 set perfect	2 sets perfect	2 sets perfect	2 sets perfect
	Alteration	Uncommon	Chlorite, etc.	Uncommon	Antigorite
	Inclusion	Uncommon	Common	Uncommon	Common
Crossed nicol	Optic character	Biaxial positive	Biaxial positive	Uniaxial negative	Biaxial positive
	Interference colour	Third order colours	Second order colours	High order pearl, gray	First order yellow
	Birefringence	0.048	0.030	0.179	0.009
	Extinction	Parallel	Symmetrical	Symmetrical	Parallel
	Twinning	Absent	Polysynthetic	Polysynthetic	Rare
	Sign of elongation	Negative	Positive	–	Positive
	Optic axial angle	84°	56°–64°	–	58°–80°

(Contd...)

Mineral	Epidote	Fayalite	Fluorite	Forsterite
Plane polarized light				
Pleochroism/colour	Pleochroic, colourless to yellowish green	Feebly pleochroic, colourless to yellowish	Nonpleochroic, colourless	Nonpleochroic, colourless
Refractive index	1.72–1.78 (high)	1.805–1.889 (high)	1.434 (low)	1.635–1.680 (high)
Shape and habit	Subhedral	Euhedral to anhedral	Euhedral or anhedral	Anhedral to subhedral
Cleavage	1 set perfect	Imperfect	2 sets perfect	Irregular fractures
Alteration	Uncommon	Grunerite	Uncommon	Antigorite
Inclusion	Uncommon	Uncommon	Uncommon	Uncommon
Crossed nicol				
Optic character	Biaxial negative	Biaxial negative	Isotropic	Biaxial positive
Interference colour	Second to third order colours	Second order colours	–	Second order colours
Birefringence	0.010–0.050	0.042–0.051	–	0.035–0.040
Extinction	Parallel	Parallel	–	Parallel
Twinning	Common	Broad lamellae	–	Absent
Sign of elongation	Positive/negative	Positive	–	Positive
Optic axial angle	69°–89°	47°–54°	–	85°–90°

Mineral	Garnet	Gibbsite	Glauconite	Glaucophane
Plane polarized light				
Pleochroism/colour	Nonpleochroic, colourless, pink	Nonpleochroic, colourless	Pleochroic, yellow to green	Pleochroic, blue to violet
Refractive index	1.741–1.887 (high)	1.554–1.589 (high)	1.59–1.644 (high)	1.621–1.668 (high)
Shape and habit	Polygonal, dodecahedral	Euhedral	Grains, pellets	Prismatic
Cleavage	Irregular fractures	Imperfect	1 set perfect	2 sets perfect
Alteration	Chlorite	Uncommon	Limonite	Uncommon
Inclusion	Common	Uncommon	Uncommon	Uncommon
Crossed nicol				
Optic character	Isotropic	Biaxial positive	Biaxial negative	Biaxial negative
Interference colour	–	Second order colours	Second order colours	Second order violet
Birefringence	–	0.022	0.020–0.032	0.013–0.018
Extinction	–	Oblique, 26°	Parallel	Oblique, 4°–6°
Twinning	–	Polysynthetic	Absent	Absent
Sign of elongation	–	Positive	Positive	Positive
Optic axial angle	–	26°	16°–30°	0°–68°

(Contd....)

Optical properties	Grunerite	Gypsum	Halite	Hedenbergite
Pleochroism/colour	Nonpleochroic, colourless	Nonpleochroic, colourless	Nonpleochroic, colourless	Nonpleochroic, colourless, greenish
Refractive index	1.657–1.717 (high)	1.520–1.529 (low)	1.544 (high)	1.732–1.757 (high)
Shape and habit	Fibrous to collumnar	Anhedral to subhedral	Anhedral	Columnar
Cleavage	2 sets perfect	1 set perfect	Perfect cubic	1 set perfect
Alteration	Uncommon	Uncommon	Uncommon	Serpentine, calcite
Inclusion	Uncommon	Uncommon	Common	Uncommon
Optic character	Biaxial negative	Biaxial positive	Isotropic	Biaxial positive
Interference colour	Second order colours	Second order yellow	–	Second order colours
Birefringence	0.042–0.054	0.009	–	0.018–0.019
Extinction	10°–15°	Parallel	–	Oblique, maximum 42°
Twinning	Polysynthetic	Polysynthetic	–	Absent
Sign of elongation	Positive	Positive or negative	–	Negative
Optic axial angle	79°–86°	58°	–	60°

(Plane polarized light — rows Pleochroism/colour through Inclusion; Crossed nicol — rows Optic character through Optic axial angle)

Optical properties	Heulandite	Hornblende	Hypersthene	Illite
Pleochroism/colour	Nonpleochroic, colourless	Pleochroic, green to brown	Pleochroic, greenish to pinkish	Pleochroic, colourless to yellowish brown
Refractive index	1.496–1.505 (low)	1.65–1.67 (high)	1.673–1.731 (high)	1.535–1.605 (high)
Shape and habit	Tabular	Prismatic	Prismatic	Irregular flakes
Cleavage	1 set perfect	2 sets perfect	2 sets	Imperfect
Alteration	Uncommon	Chlorite, epidote	Serpentine	Uncommon
Inclusion	Uncommon	Uncommon	Uncommon	Uncommon
Optic character	Biaxial positive	Biaxial negative	Biaxial negative	Biaxial negative
Interference colour	First order white	Second order colours	First order colours	Second order colours
Birefringence	0.005–0.007	0.019–0.026	0.010–0.016	0.030–0.035
Extinction	Parallel	Oblique, 12°–30°	Parallel	–
Twinning	Absent	Common	Absent	Absent
Sign of elongation	Negative	–	Positive	–
Optic axial angle	35°	52°–85°	63°–90°	10°–15°

(Plane polarized light — rows Pleochroism/colour through Inclusion; Crossed nicol — rows Optic character through Optic axial angle)

(Contd....)

	Jadeite	Kaolinite	Kyanite	Lawsonite
Mineral				
Pleochroism/colour	Pleochroic, colourless to green	Pleochroic, colourless to pale yellow	Nonpleochroic, colourless	Nonpleochroic, colourless
Refractive index	1.655–1.688 (high)	1.561–1.566 (high)	1.712–1.728 (high)	1.665–1.684 (high)
Shape and habit	Columnar, fibrous	Fine mosaic like masses	Tabular	Euhedral
Cleavage	2 sets perfect	1 set perfect	1 set perfect	2 sets
Alteration	Tremolite, actinolite	Uncommon	Clay minerals	Uncommon
Inclusion	Uncommon	Uncommon	Biotite, etc.	Uncommon
Optic character	Biaxial positive	Biaxial negative	Biaxial negative	Biaxial positive
Interference colour	Second order colours	First order white	First and second order colours	Second order blue
Birefringence	0.012–0.023	0.005–0.007	0.016	0.020
Extinction	Oblique, 30°–44°	Oblique, 1°–4°	Parallel or 7°–30°	Parallel or symmetrical
Twinning	Absent	Absent	Common	Polysynthetic
Sign of elongation	Positive	Positive	Positive/negative	Positive
Optic axial angle	70°–75°	40°	82°–83°	76°–86°

Plane polarized light rows: Pleochroism/colour, Refractive index, Shape and habit, Cleavage, Alteration, Inclusion.
Crossed nicol rows: Optic character, Interference colour, Birefringence, Extinction, Twinning, Sign of elongation, Optic axial angle.

	Lazulite	Lepidolite	Leucite	Magnesite
Mineral				
Pleochroism/colour	Pleochroic, colourless to blue	Nonpleochroic, colourless	Nonpleochroic, colourless	Nonpleochroic, colourless
Refractive index	1.603–1.642 (high)	1.530–1.590	1.508–1.509 (low)	1.50–1.70
Shape and habit	Anhedral	Tabular, prismatic	Euhedral	Subhedral
Cleavage	Imperfect	1 set perfect	Imperfect	Rhombohedral
Alteration	Uncommon	Uncommon	Uncommon	Uncommon
Inclusion	Uncommon	Uncommon	Uncommon	Uncommon
Optic character	Biaxial negative	Biaxial negative	Uniaxial positive	Uniaxial negative
Interference colour	Second and third order colours	Third order colours	First order gray	Higher order pearl gray
Birefringence	0.031–0.038	0.045	0.001	0.191–0.199
Extinction	Oblique	0°–7°	Wavy	Symmetrical
Twinning	Polysynthetic	Common	Polysynthetic	Absent
Sign of elongation	Negative	Positive	–	–
Optic axial angle	70°	40°	Small	–

(Contd...)

Mineral	Melilite	Microcline	Monazite	Montmorillonite
Plane polarized light				
Pleochroism/colour	Nonpleochroic, colourless	Nonpleochroic, colourless	Nonpleochroic, colourless to pale yellow	Pleochroic, pale pink, greenish
Refractive index	1.626–1.631 (high)	1.518–1.530 (low)	1.786–1.849 (high)	1.492–1.513 (low)
Shape and habit	Tabular	Subhedral	Small euhedral	Fine scale like
Cleavage	Imperfect	1 set perfect	Absent	Absent
Alteration	Calcite, zeolite	Clay minerals, sericite	Uncommon	Uncommon
Inclusion	Uncommon	Plagioclase	Uncommon	Uncommon
Crossed nicol				
Optic character	Uniaxial negative	Biaxial negative	Biaxial positive	Biaxial negative
Interference colour	First order gray	First order gray	Third to fourth order	Second order
Birefringence	0.005–0.006	0.007–0.010	0.050	0.021
Extinction	Parallel	Oblique, 5°–15°	Oblique, 2°–10°	–
Twinning	Absent	Cross-hatchet	Absent	Absent
Sign of elongation	–	–	–	–
Optic axial angle	–	77°–84°	6°–20°	10°–25°

Mineral	Muscovite	Natrolite	Nepheline	Olivine
Plane polarized light				
Pleochroism/colour	Nonpleochroic, colourless	Nonpleochroic, colourless	Nonpleochroic, colourless	Nonpleochroic, colourless
Refractive index	1.556–1.611 (high)	1.473–1.493 (low)	1.527–1.547	1.651–1.718 (high)
Shape and habit	Tabular	Prismatic, fibrous	Prismatic	Anhedral
Cleavage	1 set perfect	1 set perfect	Imperfect	Irregular fractures
Alteration	Clay minerals	Uncommon	Zeolite, etc.	Antigorite
Inclusion	Albite, etc.	Uncommon	In rows	Uncommon
Crossed nicol				
Optic character	Biaxial negative	Biaxial positive	Uniaxial negative	Biaxial +/–
Interference colour	First and second order	First order orange	First order gray	Second order
Birefringence	0.035–0.041	0.012	0.004	0.037–0.041
Extinction	Parallel	Parallel, symmetrical	Parallel	Parallel
Twinning	Mica law	Imperfect	Absent	Imperfect
Sign of elongation	Positive	Positive	Negative	Absent
Optic axial angle	30°–40°	60°–63°	–	70°–90°

(Contd...)

Mineral		Orthoclase	Periclase	Phlogopite	Pigeonite
Plane polarized light	Pleochroism/colour	Nonpleochroic, colourless, cloudy	Nonpleochroic, colourless	Feebly pleochroic, colourless to pale brown	Nonpleochroic, colourless
	Refractive index	1.518–1.526 (low)	1.738–1.760 (high)	1.551–1.606 (high)	1.680–1.744 (high)
	Shape and habit	Subhedral	Anhedral	Tabular, prismatic	Anhedral
	Cleavage	2 sets	Dodecahedral	1 set perfect	2 sets perfect
	Alteration	Clay minerals	Brucite	Uncommon	Serpentine, etc.
	Inclusion	Uncommon	Uncommon	Uncommon	Uncommon
Crossed nicol	Optic character	Biaxial negative	Isotropic	Biaxial negative	Biaxial positive
	Interference colour	First order gray	–	Third order colours	Second order
	Birefringence	0.007	–	0.044–0.047	0.021–0.033
	Extinction	Oblique, 5°–12°	–	Parallel to 5°	Oblique, 22°–45°
	Twinning	Simple	–	Imperfect	Polysynthetic
	Sign of elongation	Negative	–	Positive	Positive
	Optic axial angle	40°	–	0°–10°	38°–44°

Mineral		Plagioclase	Prehnite	Pyrophyllite	Quartz
Plane polarized light	Pleochroism / colour	Nonpleochroic, colourless	Nonpleochroic, colourless	Nonpleochroic, colourless	Nonpleochroic, colourless
	Refractive index	1.52–1.59	1.615–1.665 (high)	1.552–1.600 (high)	1.53–1.54 (low)
	Shape and habit	Subhedral	Tabular	Tabular	Anhedral
	Cleavage	2 sets perfect	1 set good	1 set perfect	Absent
	Alteration	Clay minerals	Uncommon	Uncommon	Absent
	Inclusion	Uncommon	Uncommon	Uncommon	Fluid, etc.
Crossed nicol	Optic character	Biaxial positive	Biaxial positive	Biaxial negative	Uniaxial positive
	Interference colour	First order gray	Second order	Third order	First order yellow
	Birefringence	0.011–0.013	0.020–0.033	0.047	0.009
	Extinction	Oblique, 0°–70°	Parallel, wavy	Parallel	Undulatory, oblique
	Twinning	Lamellar	Polysynthetic	Imperfect	Absent
	Sign of elongation	Positive/negative	Negative	Positive	Positive
	Optic axial angle	77°–78°	65°–70°	52°–62°	–

(Contd...)

Mineral		Riebeckite	Rutile	Sanidine	Scapolite
Plane polarized light	Pleochroism/colour	Pleochroic, blue to green	Pleochroic, yellowish to reddish brown	Nonpleochroic, colourless	Nonpleochroic, colourless
	Refractive index	1.693–1.697 (high)	2.603–2.903 (high)	1.517–1.526 (low)	1.54–1.607 (high)
	Shape and habit	Prismatic, fibrous	Prismatic	Euhedral	Collumnar
	Cleavage	2 sets perfect	1 set	2 sets, 1 set perfect	2 sets, 1 set perfect
	Alteration	Fibrous quartz	Uncommon	Clay minerals	Muscovite
	Inclusion	Uncommon	Uncommon	Uncommon	Uncommon
Crossed nicol	Optic character	Biaxial negative	Uniaxial positive	Biaxial negative	Uniaxial negative
	Interference colour	Masked by body colour	Deep red	First order gray, white	First and second order colours
	Birefringence	0.004	0.286–0.287	0.007	0.010–0.037
	Extinction	Inclined, 5°	Parallel	Parallel to 5°	Parallel
	Twinning	Absent	Knee shaped	Simple twin	Absent
	Sign of elongation	Negative	Positive	–	–
	Optic axial angle	80°–90°	–	0°–12°	–

Mineral		Serpentine	Sillimanite	Sodalite	Sphene
Plane polarized light	Pleochroism/colour	Pleochroic, colourless to green	Nonpleochroic, colourless	Nonpleochroic, colourless	Feebly pleochroic to nonpleochroic, colourless
	Refractive index	1.572–1.594 (high)	1.657–1.684 (high)	1.483–1.487 (low)	1.887–2.054 (high)
	Shape and habit	Anhedral	Prismatic, fibrous	Euhedral to anhedral	Euhedral, rhombic
	Cleavage	2 sets	1 set	Imperfect	Absent
	Alteration	Uncommon	Clay minerals	Zeolites	Uncommon
	Inclusion	Uncommon	Uncommon	Common	Uncommon
Crossed nicol	Optic character	Biaxial negative	Biaxial positive	Isotropic	Biaxial positive
	Interference colour	First order yellow	Second to third order colours	–	Higher order white
	Birefringence	0.007	0.020–0.023	–	0.092–0.150
	Extinction	Parallel	Parallel	–	Symmetrical
	Twinning	Absent	Absent	–	Polysynthetic
	Sign of elongation	Positive	Positive	–	–
	Optic axial angle	20°–60°	20°–30°	–	23°–50°

(Contd...)

Mineral	Spinel	Spondumen	Staurolite	Stilbite
Pleochroism/colour	Nonpleochroic, green, brown	Nonpleochroic, colourless	Pleochroic, colourles to yellow	Nonpleochroic, colourless
Refractive index	1.72–1.78 (high)	1.651–1.681 (high)	1.736–1.762 (high)	1.494–1.508 (low)
Shape and habit	Euhedral to subhedral	Tabular	Prismatic	Subhedral
Cleavage	Imperfect	2 sets perfect	Imperfect	1 set perfect
Alteration	Uncommon	Muscovite	Chlorite, etc.	Uncommon
Inclusion	Uncommon	Uncommon	Quartz	Uncommon
Optic character	Isotropic	Biaxial positive	Biaxial positive	Biaxial negative
Interference colour	–	First to second order colours	First order yellow, red	First order white
Birefringence	–	0.013–0.027	0.010–0.015	0.006–0.008
Extinction	–	Oblique, 23°–27°	Parallel	Parallel to 5°
Twinning	–	Absent	Rare	Absent
Sign of elongation	–	Positive	Positive	–
Optic axial angle	–	54°–80°	80°–88°	30°–50°

(First seven rows under *Plane polarized light*; remaining rows under *Crossed nicol*.)

Mineral	Sulphur	Talc	Topaz	Tourmaline	Tremolite
Pleochroism/colour	Pleochroic, yellow	Nonpleochroic, colourless	Nonpleochroic, colourless	Pleochroic, blue, pink, brown	Feebly pleochroic, colourless to green
Refractive index	1.95–2.24 (high)	1.538–1.590 (high)	1.607–1.638 (high)	1.62–1.67 (high)	1.60–1.655 (high)
Shape and habit	Subhedral	Fibrous	Columnar	Prismatic	Prismatic, fibrous
Cleavage	Imperfect	1 set perfect	1 set perfect	Fractures	2 sets perfect
Alteration	Uncommon	Uncommon	Uncommon	Uncommon	Talc
Inclusion	Uncommon	Uncommon	Fluid	Uncommon	Uncommon
Optic character	Biaxial positive	Biaxial negative	Biaxial positive	Uniaxial negative	Biaxial negative
Interference colour	Second order yellow	Third order colours	First order yellow	Second and third order colours	Second order colours
Birefringence	0.029	0.030–0.050	0.010	0.021–0.029	0.022–0.027
Extinction	Parallel	Parallel to 3°	Parallel to symmetrical	Parallel	Inclined, 10°–20°
Twinning	Absent	Absent	Absent	Absent	Polysynthetic
Sign of elongation	–	Positive	Negative	Negative	Positive
Optic axial angle	69°	6°–30°	48°–65°		85°–90°

(First seven rows under *Plane polarized light*; remaining rows under *Crossed nicol*.)

(Contd...)

Mineral		Tridymite	Vesuvianite	Wollastonite	Zeolite	Zircon
Plane polarized light	Pleochroism/colour	Nonpleochroic, colourless	Nonpleochroic, colourless	Nonpleochroic, colourless	Nonpleochroic, colourless	Nonpleochroic, colourless
	Refractive index	1.469–1.473 (low)	1.701–1.732 (high)	1.62–1.634 (high)	1.541–1.603 (high)	1.925–1.993 (high)
	Shape and habit	Euhedral, tabular	Euhedral, columnar	Columnar, fibrous	Subhedral	Euhedral, prismatic
	Cleavage	Absent	Imperfect	2 sets perfect	1 set perfect	Absent
	Alteration	Uncommon	Uncommon	Uncommon	Kaolin	Uncommon
	Inclusion	Uncommon	Uncommon	Uncommon	Uncommon	Common
Crossed nicol	Optic character	Biaxial positive	Uniaxial negative	Biaxial negative	Biaxial positive	Uniaxial positive
	Interference colour	First order gray	First order gray	First order orange	First order gray	Fourth order
	Birefringence	0.003	0.005	0.014	0.005–0.007	0.061
	Extinction	Inclined, undulatory	Parallel	Parallel, oblique	Parallel	Parallel
	Twinning	Absent	Absent	Present	Absent	Absent
	Sign of elongation	Positive	Negative	Positive/negative	Positive/negative	Positive
	Optic axial angle	35°–70°	–	39°	0°–76°	–

6.8 OPTICAL PROPERTIES OF ORE MINERALS IN POLISHED SECTION

Most of the ore minerals are opaque and do not allow light to pass through them. In these cases polished sections of the minerals are prepared, which are sufficiently glazed to reflect considerable amount of light. For the study of polished sections, reflecting type petrological microscope (ore microscope) is necessary. If the mineral under examination is anisotropic, the reflected light is polarised and while passing through the analyser, undergo double refraction to produce interference colour. The polished section is placed on the stage of the microscope and is first observed in plane-polarized light using the polarizer in the path of the light beam. In the second instance, the polished section is studied in cross-nicol position with both the polariser and analyser in use.

6.8.1 Study of ore Minerals in Plane Polarised Light

In plane-polarised light, properties like nature of polish, colour, pleochroism, reflectivity, form and habit, cleavage and polishing hardness are studied.

6.8.1.1 Nature of Polish

Different minerals have different internal atomic configurations and thus behave differently when polished sections are prepared. Some of the minerals take very good polish while others take bad polish. This criterion helps in identification of many ore minerals.

Commonly galena shows consistent triangular pits due to intersection of cleavages while these pits are rare in pentlandite, wolframite, lorandite, franklinite, jacobsite, clausthalite, etc.

6.8.1.2 Shape and Habit

Most metallic minerals are xenomorphic in polished section, i.e. they have no definite form. Generally the harder minerals and those having higher melting temperature, have a tendency to develop their own crystallographic habits (idiomorphic). Pyrite, arsenopyrite, sperrylite, magnetite, native osmium, cassiterite, eskolaite, hematite, wolframite, melonite, chromite are some of the examples. Magnetite, arsenopyrite and pyrite show octahedral, prismatic and cubic habits respectively. Euhedral arsenopyrite and eskolaite are shown in Fig. 6.17. Crystal habits offer valuable clues for identification of minerals and are of paramount importance in paragenetic study.

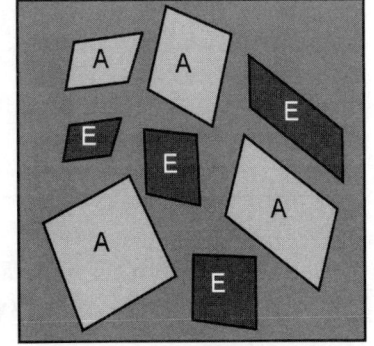

Fig. 6.17: Euhedral crystals of arsenopyrite (A) and eskolaite(E)

The euhedral character accentuates the substitution tendency of other minerals of the paragenesis sequence, which is useful in studying the succession of crystallisation of the minerals. Some minerals formed at low temperature show colloform texture with colloidal features. They occur in form of concentric layers with convex or spherulitic opal-like texture (Fig. 6.18). Goethite, schalenblende, melnicovite, pitchende, psilomelane are good examples.

Skeletal texture (Fig. 6.19) is formed occasionally due to incomplete development of crystal habit. The mineral exhibiting skeletal texture shows a euhedral tendency but is closely associated with other minerals suggesting replacement. This is particularly the cases of magnetite and ilmenite in mafic volcanic rocks and, sometimes, of sphalerite or pyrite in

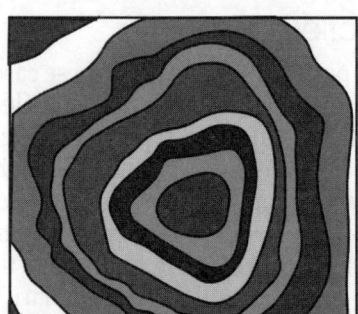

Fig. 6.18: Colloform texture of goethite

Fig. 6.19: Skeletal texture of ilmenite in mafic rock

volcano-sedimentary deposits. In some cases, this type of texture is formed due to rapid crystallization in a cooled or abruptly supersaturated environment.

6.8.1.3 Pleochroism

It is the property of certain minerals exhibiting variation of colour along with variation of the reflectance during rotation of the stage of the microscope. Some minerals show very striking pleochroism. Most minerals, however, exhibit relatively distinct to faint and sometimes very low pleochroism. Degrees of pleochroism shown by some minerals are given in Table 6.5.

Table 6.5: Degree of pleochroism shown by different ore minerals

Degree of pleochroism	Minerals
Very strong	Covellite
Strong	Ilmenite, magnetite, psilomelane (excluding fine grained varieties)
Moderate	Pyrrhotite, pyrolusite (excluding fine grained varieties)
Weak	Hematite, arsenopyrite, chalcocite, chalcopyrite, braunite

As a rule, all isotropic minerals are nonpleochroic, whereas all anisotropic minerals are not necessarily pleochroic. Anisotropic minerals show pleochroism in varying degrees.

6.8.1.4 Colour

The colour of the mineral attracts the attention of the observer. More commonly, metallic minerals are either indistinctly coloured (white or gray with pale cream or bluish shades) or colourless. In some cases, the colour may be strong, as in the case of most copper minerals, particularly sulfides. Some minerals show different colours according to their association with other minerals. For example, wittichenite is grayish brown adjacent to chalcopyrite, distinctly white with bornite and grayish brown with purple tints when associated with emplectite or bismuthinite. Pyrite, which is pale yellowish white, turns greenish gray with native gold and yellow when borders galena. Thus, association should be taken into account before attributing any particular shade to a mineral. Some of the most common colours of minerals are given in Table 6.6.

Table 6.6: Characteristic colours of different minerals

Colour	Minerals
Blue	Chalcocite, covellite, psilomelane, sphalerite (grayish), hematite
Bluish	Pyrargyrite, proustite, samsonite, cuprite
Brown	Pyrrhotite, eskebornite, enargite, stannoidite, cubanite
Brownish	Native selenium, tenorite, hakite, ilmenite, berthierite, seligmannite
Cream	Bismuth, linneite, bravoite, pentlandite, penroseite, bornhardtite
Greenish	Stannite, tetrahedrite-tennantite, bournonite, argentite
Gray	Sphalerite (brownish tint), magnetite (brownish tint), psilomelane (bluish tint)
Orange	Nickeline, renierite, luzonite, mawsonite, colusite, genkinite
Pink	Pyrrhotite, bornite, ilmenite, chromite, sphalerite, magnetite,
Pinkish	Cobaltite, maucherite, kotulskite, chalcothallite, rohaite
Purple	Germanite, bornite (when tarnished), umangite, villamaninite
Red	Native copper, rickardite (pinkish), koutekite (when etched)
White	Galena (with brownish tint), hematite (with bluish tint), pyrolusite (with yellowish tint), pyrite (in association with chalcopyrite), chalcocite (in association with covellite), arsenopyrite (with creamy tint)
Yellow	Pyrite, chalcopyrite, pyrolusite, native gold
Yellowish	Pyrite, millerite, makinenite, heazlewoodite, lazarevicite, sulvanite

In oil immersion, the intensity of colour generally increases. In case of strongly coloured minerals, such as covellite, the variation of colour produced by immersion is striking, i.e. from blue to purplish red.

6.8.1.5 Reflectivity

This refers to the reflecting power of a mineral and is expressed as the percentage of incident light reflected from the surface of polished section. Reflectance has always been considered as the primary criterion in the identification of metallic mineral. In case of anisotropic minerals, which are pleochroic, reflectance varies considerably with the grain under examination and depends on the crystallographic orientation. Reflectance depends on the refractive index and the coefficient of absorption of light by the mineral. In case of metals and minerals with metallic lustre, most of the light is reflected and therefore, the reflectance is close to 100%. On the other hand, in the case of gangue minerals, which are transparent in thin section, greater part of incident light penetrates and is absorbed by the mineral, and therefore, only a small part of it is reflected back. In these cases the reflectance is closer to 0%. Due to these reasons, every transparent mineral in thin section seems very dark in polished section. Reflectance values can be measured and quantified by means of a photoelectric cell or an electron photo-multiplier. Frequently, the reflectivity power of the ore mineral is expressed qualitatively as low, moderate, high and very high. Reflectivities of some common minerals are given in Table 6.7.

6.8.1.6 Cleavage

It is an important criterion for mineral identification. Some minerals have numerous and perfect cleavages which on overlapping, form triangular pits during polishing. These pits, in the form of black aligned triangles on the polished section, are excellent criteria of identification,

Table 6.7: Reflectivity values (in percent) of some common minerals

Degree of reflectivity	Minerals
Very high	Aresenopyrite (53), pyrite (54), galena (43.5)
High	Chalcopyrite (41.5), pyrrhotite (37), pyrolusite (31–42)
Moderate	Hematite (26), chalcocite (22.5–30), covellite (27), psilomelane (28–33).
Low	Magnetite (21), ilmenite (18–20), covellite (18.5), bornite (18.5), sphalerite (18.5), chromite (15).

especially in case of galena. Some minerals, such as molybdenite, graphite, valleriite, etc. show mica-like cleavages. Cleavages may be perfect, imperfect or distinct. Some minerals showing these types of cleavages are given in Table 6.8.

Table 6.8: Cleavages of some common minerals

Type of cleavage	Minerals
Perfect cleavage	Arsenolamprite, bismuthinite, eskebornite, chalcothallite, covellite, franckeite, graphite, molybdenite, sternbergite, tellurobismutite, tetradymite, tungstenite, valleriite
Imperfect cleavage	Athabascaite, bukovite, crednerite, franklinite, galena, hollandite, laitakarite, livingstonite, lorandite, manganite, orpiment, patronite, pentlandite, pyrolusite, quenselite, siderite, smithite, stibnite, teallite, zincite
Distinct cleavage	Argentopyrite, chalcocite, corvusite, digenite, hausmannite, jacobsite, jamesonite, kobellite, krennerite, magnetoplumbite, manganosite, plumboferrite, realgar, selenium, semseyite, sulvanite, ullmannite, weibullite, wittite

6.8.1.7 Polishing hardness

The hardness of a mineral is tested under the microscope by scratching the surface by a sharp steel needle with a weighted handle. The minerals scratched under the weight of the handle are said to be soft, with moderate pressure are intermediate and those scratched with relatively higher pressure are grouped under hard and very hard categories.

Two minerals of similar hardness show very fine boundaries because there is no difference of relief between them. However, when one mineral is harder than another adjacent mineral, polishing wears the soft mineral more while the harder mineral stands up in relief. Hence, the boundary between two minerals of different hardness seems coarse and blurred. This characteristic of hardness is often employed for the identification of some metallic minerals. Employing the hardness criteria, cubanite is distinguished from pyrrhotite, though both are commonly associated with chalcopyrite. Both are brownish and have more or less the same reflectance, but the hardness of cubanite is similar to chalcopyrite and the boundaries between the two are very indistinct. On the other hand, pyrrhotite is harder than chalcopyrite and stands out in relief with respect to chalcopyrite showing much coarser boundaries.

A practical way to determine the relative hardness of two minerals is to examine the polishing scratches. Very soft minerals such as argentite and stromeyerite are always quite noticeably scratched in comparison to others.

A microhardness tester is a microscope in which one of the objectives is replaced by a diamond octahedron oriented downward. A shutter release applies the load of the diamond on the grain previously centered with a normal objective and a cross-hair ocular. The diamond leaves a square indentation on the section. The diagonal of the square indicates the hardness of the mineral, which is known as Vickers hardness. This device is often used to differentiate pyrite from pentlandite. With this method, the harder pyrite shows tiny and indistinct indentations while softer pentlandite gives deeper and almost perfect notch.

6.8.2 Study of Ore Minerals under Crossed Nicols

Optical properties like isotropism/anisotropism, interference (polarisation) colour, twinning, zoning, internal reflection and etch-test are studied in crossed nicol position.

6.8.2.1 Optic Character (Isotropism/Anisotropism)

Minerals crystallising in isometric system are isotropic as the vibration direction of light leaving polariser remains unchanged and is discarded by the analyser resulting in darkness of the mineral. Minerals crystallising in other systems are anisotropic. The degree of anisotropism is completely independent of the colours. Highly anisotropic minerals may show coloured anisotropism as in case of covellite or be completely colourless (chalcophanite). Weakly anisotropic minerals often have slight chromatic effect, but this is not always true. For example, braunite polarises very weakly and with no tint, whereas gallite, with weak anisotropism, polarises in tints ranging from reddish to greenish colours (with uncrossed nicols). Although independent of the colour, anisotropic intensity cannot be dissociated from it. The intensity and anisotropism of colours may be determined qualitatively by sight. Some of the minerals of different degrees of anisotropism are listed in Table 6.9.

Table 6.9: Degree of anisotropism of ore minerals	
Degree of anisotropism	*Minerals*
Very strong	Covellite
Strong	Psilomelane, ilmenite
Moderate	Pyrrhotite, pyrolucite
Weak	Chalcopyrite, chalcocite, hematite, arsenopyrite

6.8.2.2 Interference Colour

Interference colours are considered as one of the best criteria for identification of ore minerals. In contrast to transparent minerals, all ore minerals do not show their most characteristic colours with perfectly crossed nicols. If the nicols are uncrossed to certain extent, mawsonite shows tints ranging from yellow to very bright green. Similarly, millerite shows tints varying from bright yellow to blue, only with slightly uncrossed nicols. To certain extent, the interference colours depend on the objective used. For example, with a low-power objective (5×), eucairite shows violet-blue to green interference colour, whereas with a medium-power objective (16×) the colour ranges from brick-red to green.

6.8.2.3 Twinning

The ore minerals commonly show contact, interpenetrant, polysynthetic twins. Mineral like safflorite shows contact twins of six orthorhombic crystals, which appear as an impressive six-pointed star twin (Fig. 6.20). Interpenetrant twins are rare but are highly characteristic. Generally, four-fold interpenetrant twin (Fig. 6.21) is pseudohexagonal. It is seen in marcasite, argentopyrite, picotpaulite and stibiopalladinite. Polysynthetic twins (Fig. 6.22) are seen as alternate lamellae similar to plagioclase lamellae in thin section. They either may be only in one direction as in diaphorite, or in several directions as in rathite, bournonite and parkerite. They may show parallel orientation (baumhauerite) or wedge-shaped (ramdohrite). Sometimes, they may be twisted due to mineral deformation, as in stibnite and bismuthinite. Some examples of twins are given in Table 6.10.

Fig. 6.20: Six-pointed star twin of safflorite

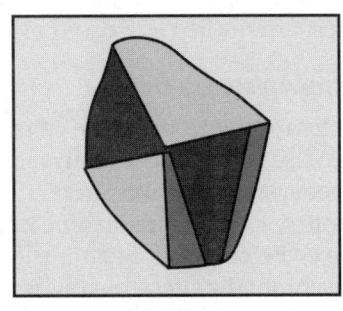

Fig. 6.21: Interpenetrant twin of picotpaulite

Fig. 6.22: Polysynthetic twin of baumhauerite

Table 6.10: Twinning of ore minerals

Type of twinning	Minerals
Complex twins	Cobaltite, rhodostannite, arsenopyrite
Four-fold interpenetrant twins	Argentopyrite, picotpaulite, marcasite, dyscrasite, stibiopalladinite, geocronite, godlevskite, cubanite
Polysynthetic in one direction	Argentite, chalcocite, marcasite, metacinnabar, stephanite, sylvanite twinnite
Polysynthetic in several directions	Antimony, arsenic, bismuth, hausmannite, tenorite, hematite, hessite, ilmenite, hollandite, rathite
Serrated twins	Native arsenic
Simple twins	Cassiterite, ferroselite, hastite, renierite, gudmundite, roquesite, marcasite
Sinuous twins	Stibnite, bismuthinite, orpiment, livingstonite, cinnabar, franckeite, pyrrhotite, bukovite, chalcostibite, chalcothallite
Six-pointed star twins	Arsenopyrite, safflorite, lingite
Wedge-shaped polysynthetic twins	Ramdohrite, andorite, crednerite, bismuth, cuprostibite

6.8.2.4 Zoning

The compositional variation of the mineral expressed by sequential changes in optical character, particularly difference in anisotropism is termed zoning. It is generally common in case of euhedral minerals like galena, magnetite, arsenopyrite, bravoite, etc. (Fig. 6.23).

Fig. 6.23: Zoning in ore mineral

6.8.2.5 Internal Reflection

Some ore minerals are not completely opaque under crossed nicols. Some amount of light enters in and is reflected from the cleavage planes, twin planes and grain boundaries. This phenomenon is known as internal reflection. The colour of the internally reflected light is the characteristic of the concerned mineral, which is used as identification criteria. Internal reflection is extensive in realgar, proustite and smithite and rare in pyrargyrite, polybasite, miargyrite, etc. Characteristic internal reflections are given in Table 6.11.

Table 6.11: Characteristic internal reflection of ore minerals

Colour of internal reflection	Mineral
Blue	Azurite
Brown	Sphalerite, cassiterite, galaxite, pyrochlore, perovskite, chromite, uraninite
Colorless	Scheelite
Green	Malachite, eskolaite, manganosite
Red	Hematite and ruby silver
Yellow	Orpiment, sphalerite, cassiterite, wurtzite, goethite

6.8.2.6 Etch-test

If a drop of chemically active solution is placed on a polished surface, chemical reaction takes place. The solution affects different minerals to different degrees producing characteristic tarnish. Some of the common solutions are HNO_3, HCl, KCN, $FeCl_3$, KOH, $HgCl_2$, H_2O_2, aqua regia, etc. This criterion is used for identification of ore minerals in certain cases. Though it is not advisable to use corrosive reagent because it destroys the polished section, in some ambiguous cases, this method becomes essential. For example, hydrogen peroxide is used to distinguish between potassic and barytic psilomelane. Hydrochloric acid etching can help in distinguishing between platinum and palladium. Different minerals respond differently to etch test. Minerals like polybasite, pearceite, argentite, jalpaite, argyrodite and canfieldite show quick reaction while stromeyerite and stephanite react very slowly with the etching reagent. Even in air, arsenic, bornite and talnakhite oxidise rapidly while dyscrasite, aurostibite, novakite, koutekite, paxite, maldonite, native bismuth, native copper, berthierite, alabandite, etc. oxidise at a relatively slower rate.

6.8.2.7 Miscellaneous Criteria

There are certain criteria, which are not based on the optical characteristics of the mineral but are useful in mineral identification. These are exsolution, replacement textures, mineral associations and paragenesis.

   i. **Exsolution:** Crystallization of components of a solid solution series following super-saturation or unmixing below a certain temperature is known as exsolution. In exsolution textures, a minor mineral is included within the more abundant mineral. The exsolution textures are very characteristic and may occur in the form of blebs (Fig. 6.24), lamellae (Fig. 6.25), flames (Fig. 6.26), star (Fig. 6.27), myrmekitic (Fig. 6.28), etc. Examples of exolution are given in Table 6.12.

   ii. **Replacement:** Replacement refers to substitution of one mineral by one or several other minerals by phenomena different from exsolution but giving rise to similar textures. For instance, the replacement of magnetite by hematite or manganosite by hausmannite occurs as lamellae related to the cleavage of the replaced mineral. Similarly, bornite may be replaced by chalcopyrite. Replacement of cassiterite crystal by galena is shown in Fig. 6.29. Generally replacement textures are not as regular as are exsolution textures. The replacement of bornite by chalcocite may lead to myrmekitic textures while replacement of sphalerite by chalcopyrite occurs in form of blebs.

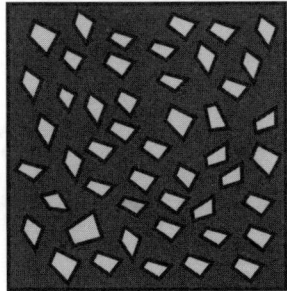

Fig. 6.24: Blebs of chalcopyrite in sphalerite

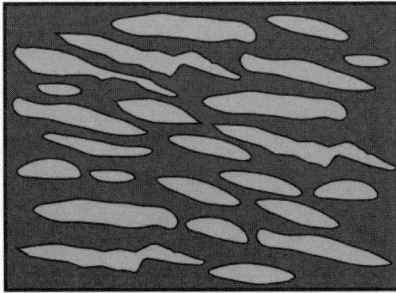

Fig. 6.25: Lamellae of chalcopyrite in briartite

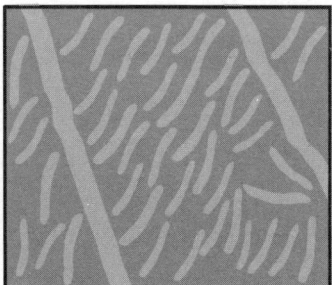

Fig. 6.26: Flames of pentlandite in pyrrhotite

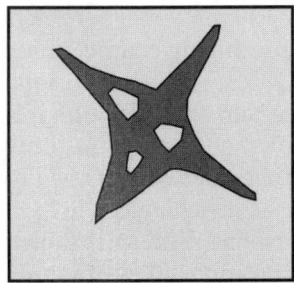

Fig. 6.27: Star like exsolution of sphalerite in chalcopyrite

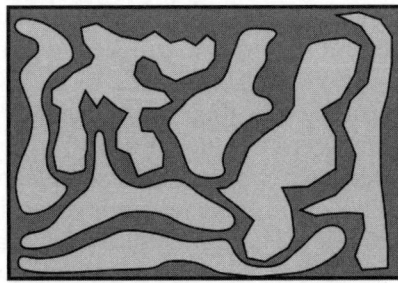

Fig. 6.28: Myrmekitic association of chalcopyrite and bornite

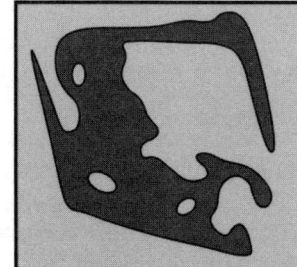

Fig. 6.29: Replacement of cassiterite by galena

Table 6.12: Examples of exsolution of ore minerals

Type of exsolution	Mineral
Blebs	Chalcopyrite, pyrrhotite or stannite in sphalerite, sphalerite and chalcopyrite in stannite, etc.
Flame-like	Pentlandite in pyrrhotite, mackinawite in chalcopyrite and pentlandite
Lamellae	Ilmenite in magnetite, cubanite in chalcopyrite, chalcopyrite and chalcocite in bornite, galena in tetrahedrite-tennantite, polybasite, bournonite and pyrargyrite in galena, ilmenite in hematite, hematite in ilmenite, hausmannite in zincite, hematite in cassiterite, etc
Myrmekitic	Arsenic- antimony-stibarsen, magnetite-ulvite, realgar- stibnite, etc.
Star-shaped	Sphalerite and stannite in chalcopyrite

Some minerals have the tendency to develop dendritic textures in the gangue or other minerals (Fig. 6.30). Many cobalt-nickel arsenides develop dendrites with native bismuth or silver. The latter often occur as rows of small cubic crystals incompatible with their own habit. Thus, they seem to have replaced pre-existing cubic crystals of native silver. In some cases, replacement may give rise to hollow crystals (Fig. 6.31). If the replacement of outer form is complete, pseudomorph of one mineral after another may result (Fig. 6.32). Replacement of organic tissue by pyrite (Fig. 6.33) is a common phenomenon in many coal seams.

iii. **Mineral associations—parageneses:** Paragenesis refers to sequence of crystallization of minerals. In the process, some minerals are associated with certain minerals in preference to others. Some of the metallic elements have strong affinities for each other. For example, nickel and chromium are generally found in the same deposit. This is also true for cobalt,

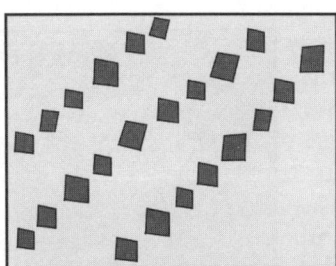

Fig. 6.30: Rows of proustatite cubes in rammelsbergite

Fig. 6.31: Formation of hollow crystals by replacement

Fig. 6.32: Pseudomorph of galena after pyromorphite

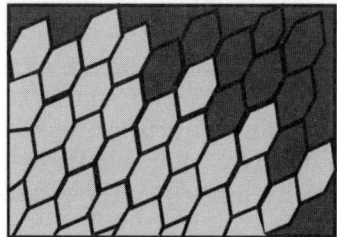

Fig. 6.33: Replacement of organic tissue by pyrite

nickel, bismuth and silver; lead, zinc and silver; silver, zinc and antimony; tungsten, tin and molybdenum, etc. Minerals containing these elements also show the same affinities. This feature is very important and helps in determining various minerals of a paragenesis. If the identification of a metallic mineral within an association is found to be difficult, the identification of adjacent minerals within the same paragenesis will provide valuable clues for its identification.

6.8.2.8 Ore Texture

Ore texture refers to the size, shape and arrangement of one or more ore minerals, which make up the bulk of ore. Though some of the textures can be seen in megascopic specimens, most of them are best seen under petrological microscope. Different types of textures developed in various situations are described below.

i. **Dendrites:** These are branched skeletal crystals developed at high temperature. The branches have parabolic tips with secondary and tertiary branching. Linear arrangement of magnetite and ilmenite in suddenly cooled basalts are good examples.

ii. **Isolated equant euhedral rounded (near spherical) crystals:** From oxide-ore rich melts magnetite, ilmenite, chromite, pyrrhotite crystallise simultaneously that leads to the interference of growth. As a result, the crystals form clustering of equant grains.

iii. **Immiscible sulphide droplets in silicate matrix:** Spherical droplets that join to form *dumbbell-shaped* structures are formed when immiscible sulphide globules separate out from mafic magmas.

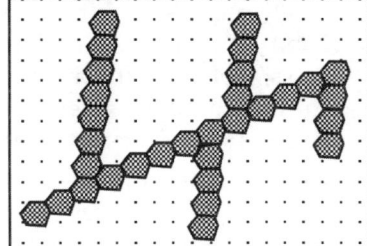

Fig. 6.34: Multiple chain texture

iv. **Synneusis or swimming together:** This type of texture develops due to simultaneous crystallization of mutually interfering chromite grains with widely varying interfacial angles. It is similar to welding-together texture. Sometimes, numerous chromite grains forming networks appear to swim together in a pool of silicate matrix forming *multiple chain texture* (Fig. 6.34). At times, centrally located silicate minerals looking like a lagoon are surrounded by chromite grains forming *atoll texture* (Fig. 6.35). In many instances, a mosaic of chromite grains with interstices filled with olivine or pyroxene or their altered products like serpentine, talc, etc. occurs. In such case the texture is known as *cumulus texture*, where the chromite occurs as cumulate (Fig. 6.36). In such cases, the chromite grains often become more or less round due to the corrosion leading to the formation of *clot texture* or *deuteric texture* (Fig. 6.37).

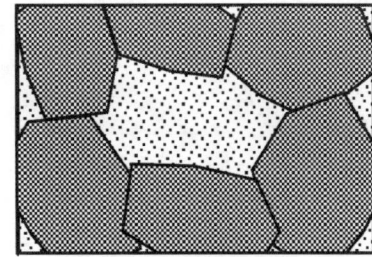

Fig. 6.35: Atoll texture

v. **Orbicular texture:** This type of texture consists of concentric shells of chromite and olivine forming

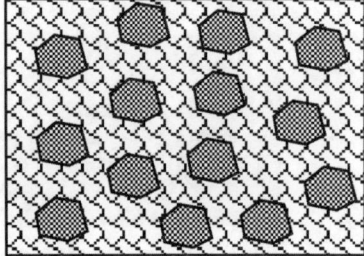

Fig. 6.36: Cumulus texture

spheroidal masses of 2–15 cm diameter. The term *nodular texture* is applied when the individual nodules are composed of coarser chromite grains of several centimeters in diameter.

vi. **Spheroidal clots:** This type of texture is formed when fine-grained chromite grains settle down to the bottom of magma chamber, where hindrances inhibit their growth.

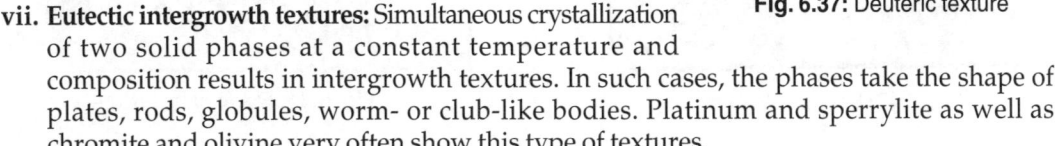

Fig. 6.37: Deuteric texture

vii. **Eutectic intergrowth textures:** Simultaneous crystallization of two solid phases at a constant temperature and composition results in intergrowth textures. In such cases, the phases take the shape of plates, rods, globules, worm- or club-like bodies. Platinum and sperrylite as well as chromite and olivine very often show this type of textures.

viii. **Zoning:** When the dynamic equilibrium with the changing composition cannot be maintained, crystals with variable compositional layers (zoned crystals) are formed (Fig. 6.23).

ix. **Comb texture:** In fissures and cavities, ore fluid with low degree of supersaturation produces few nuclei, which get a chance to form larger crystals. In such cases, crystal growth starts from the confining wall and project into the fluid-filled cavity or fissure. In the process, overcrowding and impingement of all growing crystals take place. Only those crystals whose directions of fastest growth are favorably oriented (e.g. the major diagonal of arsenopyrite rhomb) survive, while others are suppressed. The texture formed in this process is termed comb texture (Fig. 6.38).

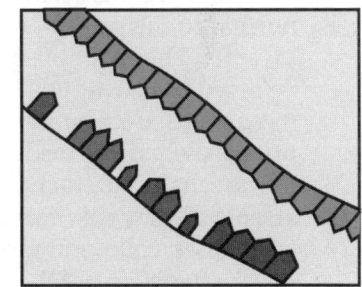

Fig. 6.38: Comb texture

x. **Crustification banding:** While forming the comb texture the chemistry of the fluid may change resulting in production of successive layers of different minerals. The resulting texture is known as crustification banding. Such bands of different composition may be symmetrically or asymmetrically disposed on either side of a median fissure region (Fig. 6.39).

xi. **Cockade texture:** When the ore-fluid is moderately supersaturated, large number of crystals of relatively smaller grain size is formed. This leads to the formation of cockade texture in which concentric growth of different mineral layers around brecciated fragments take place (Fig. 6.40). Sometimes, intramineralisation leads to brecciation and fracturing.

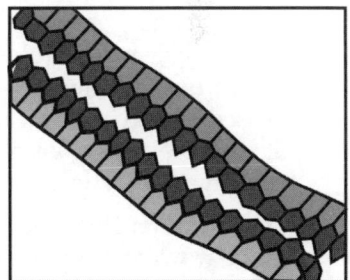

Fig. 6.39: Crustified band

xii. **Framboidal texture:** It is nearly spherical close-packed aggregates of minute euhedral sulfide crystals commonly of pyrite (Fig. 6.41). It is so named for its resemblance to bunch of raspberry. Framboids made up of chalcopyrite, limonite and pyrolusite are also known in which replacement of pyrite is well visualised. Larger aggregations of such spheres may fill shells of foraminiferids or give rise to polyframboids and nodules.

Fig. 6.40: Cockade texture

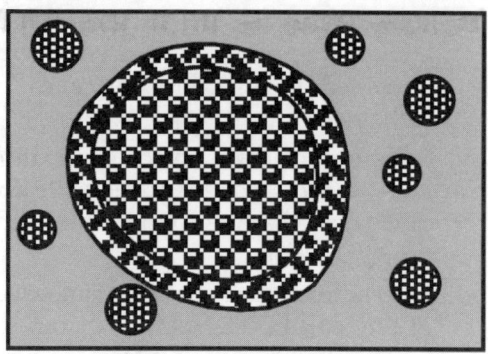

Fig. 6.41: Framboidal texture

xiii. Colloform texture: Highly oversaturated solutions can give rise to large number of nuclei. In some cases, even colloidal gels may result. The banded structures are characterized by concentric zoning normal to the growth direction of constituent crystals. The reaction-diffusion mechanism is believed to be responsible for the formation of these bands. Cobalt-nickel arsenides, pitchblende, manganese and iron oxide and hydroxide minerals, pyrite, chalcocite including native copper and arsenic display colloform textures (Fig. 6.18). In some cases, colloform texture resembles the fish-scale pattern (Fig. 6.42).

Fig. 6.42: Fish-scale type colloform texture

xiv. Ooliths (0.25–2.0 mm), pisoliths (> 2.0 mm) and spastoliths (deformed pisoliths): These spherical bodies are characteristic textural features of manganese ores. These textures may be accretionary, concretionary or microbial. Accretion of particles leads to the formation of rounded shape while concentric layering develops due to rolling on a substrate or above the sediment/water interface. Ooliths formed by this process are always synsedimentary. On the other hand, concretions are produced by chemical precipitation at multiple centers within the sediment during or after diagenesis. Such textures with confirmed affiliation to microbiological colonies have been termed *oncolites*.

xv. Foam texture: This type of texture develops when grain-growth takes place in a rigid medium. Since voids are not available for crystal growth in the rigid matrix, the crystal growth can no longer be achieved by spherical forms but through development of the foam texture (Fig. 6.43) that consists of juxtaposition of mineral grains forming triple point junctions at 120°, as in case of chromite. Coarsening of grain size concurrently causes reduction in surface area and consequent reduction of interfacial energy. For aggregates of two or

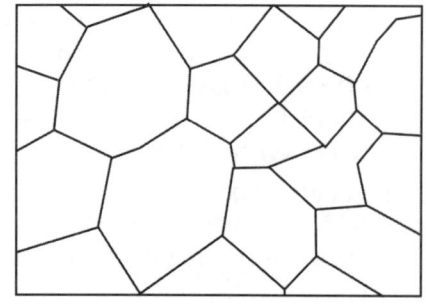

Fig. 6.43: Foam texture

more minerals, the angles at the junction between grains of different minerals (dihedral angle) are variable depending on the surface energies of the phases involved. The characteristic dihedral angle between the galena-sphalerite pair is 130°.

xvi. **Exsolution:** Unmixing and decomposition textures develop from subsolidus precipitation of new phases. Growth from numerous independent nuclei results in the formation of blebs (Fig. 6.24), lamellar (Fig. 6.25), flame (Fig. 6.26), star like (Fig. 6.27) and myrmekite associations (Fig. 6.28).

xvii. **Inversion twins:** Inversion twins are expression of strain associated with phase change in solid state. In cubic-tetragonal inversion of chalcopyrites on cooling, transformation from higher to low symmetry phase involves movement and redistribution of atoms. Rise of temperature generates thermal twins. Unlike coherent growth-twins, the thermal twin interfaces are of higher energy planes and are easily recognized by etch-test.

xviii. **Reaction rims:** These are confined to grain boundaries between specific mineral pairs. They reflect subsolidus reaction and the tendency to sustain equilibrium under newly imposed thermal conditions.

xix. **Thermal stress fracture:** It is a special feature of pentlandite grains, which have cooled at an elevated temperature. The unusually high coefficient of thermal expansion (2–10 times higher than those of common sulfides) of pentlandite results in this typical textural feature.

xx. **Replacement texture:** Replacement by substitution as well as replacement by decomposition or dissolution are both possible. Typical growth textures vary in scale from megascopic to microscopic and include features that are indicative of syn- and post-depositional and/or post-lithification precipitation/redistribution of ore minerals. Replacement may result in the formation of hollow (Fig. 6.31), incipient (Fig. 6.32) or complete crystals giving rise to pseudomorph of one mineral after another (*see* Fig. 6.32).

xxi. **Deformed textures:** Gneissosity, schistosity and grain flattening and elongation have been observed in case of minerals like galena, chalcopyrite and pyrrhotite in many deformed ores. Preferred orientation of sphalerite deformation twins on (111) forming at 30°–40° to the maximum compression axis has been demonstrated in oriented natural ore specimens and by experimental deformation studies. Orientation of pyrrhotite (0001) plane parallel to ore body banding has been reported from several metamorphosed ore bodies. Syntectonic porphyroblasts of pyrite, pyrrhotite and arsenopyrite (Fig. 6.44) are frequently preserved in metamorphosed ores. Curved inclusion trails and/or fine-grained inclusions still retaining the orientation and grain-size they had before being captured, provide clues to the syntectonic origin of the porphyroblasts. Por-

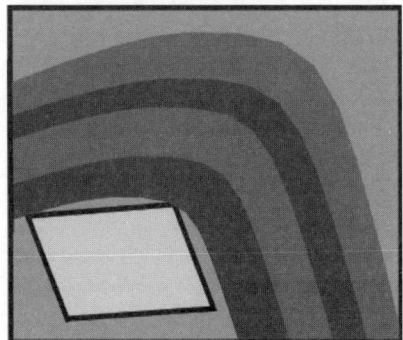

Fig. 6.44: Porphyroblast of arsenopyrite protruding into a sphalerite grain. Note the bending of twin lamellae of sphalerite

phyroclasts of shattered pyrite, magnetite and sphalerite stringing out in recrystallised galena, chalcopyrite or pyrrhotite matrixes are common.

Optical properties of some common ore minerals are given in Table 6.13.

Table 6.13: Optical properties of ore minerals

	Mineral	Argentite	Arsenopyrite	Azurite	Bornite	Braunite
Plane polarized light	Nature of polish	Poor, scratches common	Excellent	Good	Good	Good
	Shape and habit	Euhedral to subhedral	Euhedral, lozenge-shaped	Crystalline with colloform bands	Euhedral	Euhedral to subhedral
	Pleochroism	Weak	Weak	Strong	Nonpleochroic	Pleochroic
	Colour	Light gray with a greenish shade	White with a very slight yellowish tinge	Gray	Pinkish brown, orange	Gray to brownish
	Reflectivity	Low	Very high	Low	Low	Low
	Cleavage	Present	Absent	Imperfect, fractures present	2 sets	Absent
	Polishing hardness	Low	High	Low	Low	High
Crossed nicol	Optic character	Anisotropic	Anisotropic	Anisotropic	Weakly anisotropic	Weakly anisotropic
	Interference colour	Purplish blue-green to green	Blue to brown	Blue	Brown-gray to pink-brown	Gray to brownish gray
	Twinning	Lamellar	Hourglass-shaped	Rare	Polysynthetic	Rare
	Zoning	Absent	Common	Absent	Absent	May be present
	Internal reflection	Absent	Bluish green to reddish brown	Deep blue	–	Absent
	Etch test	–	HgCl$_2$ tarnishes brown	Negative	KCN stains brown	SnCl$_2$+HCl stain brown

	Mineral	Chalcocite	Chalcopyrite	Chromite	Cinnabar
Plane polarized light	Nature of polish	Good	Good	Good	Good
	Shape and habit	Subhedral	Anhedral	Euhedral, octahedral	Anhedral
	Pleochroism	Nonpleochroic	Nonpleochroic	Nonpleochroic	Pleochroic
	Colour	White or gray-white with a bluish shade	Pale yellow to brass yellow	Gray with brownish tint	Yellowish gray-white to brownish gray-white
	Reflectivity	Moderate	High	Low	Moderate
	Cleavage	Imperfect	Imperfect	Imperfect, fractures present	Imperfect
	Polishing hardness	Low	Low	High	Moderate

(Contd...)

	Property				
Crossed nicol	Optic character	Weakly anisotropic	Very weakly anisotropic	Isotropic	Strongly anisotropic
	Interference colour	Orange-brown	Greenish yellow	–	Light green
	Twinning	Lamellar	Lamellar	Absent	Polysynthetic
	Zoning	Absent	Absent	Occasional	–
	Internal reflection	–	–	Red brown	Red
	Etch test	Corrosion with HNO_3	Aqua regia stains brown	Negative	–

	Mineral	Cobaltite	Covellite	Cuprite	Galena
Plane polarized light	Nature of polish	Poor	Good	Poor, triangular pits present	Poor, triangular pits present
	Shape and habit	Euhedral	Anhedral to subhedral	Anhedral to subhedral	Euhedral, squarish
	Pleochroism	Pleochroic	Pleochroic	Nonpleochroic	Nonpleochroic
	Colour	White with a faint pink shade	Blue to bluish white	Gray-white with a bluish tint	White
	Reflectivity	High	Moderate	Moderate	High
	Cleavage	Imperfect	Distinct, basal	Imperfect	3 sets perfect with triangular pit
	Polishing hardness	High	Low	High	Low
Crossed nicol	Optic character	Weakly anisotropic	Anisotropic	Anisotropic	Isotropic
	Interference colour	Yellow-brown to bluish gray	Red to orange	Green to blue tints	–
	Twinning	Complex twins with lamellar intergrowths	Twisted lamellae	Indistinct	Indistinct
	Zoning	Absent	Absent	Absent	Absent
	Internal reflection	Nil	Violet to pink	Red	Absent
	Etch test	–	KCN stains red or black	–	HNO_3 stains black

(Contd...)

Mineral		Goethite	Hausmannite	Hematite	Ilmenite	Lithiphorite
Plane polarized light	Nature of polish	Good	Good	Bad	Excellent	Good
	Shape and habit	Botryoidal, mammilliform	Euhedral	Tabular, rhombohedral	Tabular, rhombohedral	Radiating fibers
	Pleochroism	Weakly pleochroic	Pleochroic	Weakly pleochroic	Pleochroic	Pleochroic
	Colour	Gray-white with a bluish tint	Gray to graywhite	Grayish white-with bluish tint	Gray with brownish shade	White to gray
	Reflectivity	Low to moderate	Low	Moderate	Low	Moderate
	Cleavage	Imperfect	Imperfect	Imperfect, parting present	Imperfect, parting present	Imperfect
	Polishing hardness	High	High	Very high	High	Low
Crossed nicol	Optic character	Anisotropic, colloform varieties are isotropic	Anisotropic	Anisotropic	Anisotropic	Anisotropic
	Interference colour	Red, orange or yellowish	White to bluish gray	Greenish gray to light brownish	Greenish gray	White with a bluish shade
	Twinning	–	Lamellar	Lamellar	Lamellar	–
	Zoning	Present	Absent	Absent	Absent	Absent
	Internal reflection	Brown	Red	Deep red	Dark brown	Not seen
	Etch test	Negative	Negative	Negative	Negative	–

Mineral		Magnetite	Malachite	Manganite	Marcasite	Molybdenite
Plane polarized light	Nature of polish	Good	Good	Good	Good	Poor
	Shape and habit	Euhedral, octahedral	Cryptocrystalline with concentric growth bands	Prismatic, columnar	Colloform aggregates with radiating texture	Cryptocrystalline
	Pleochroism	Nonpleochroic	Pleochroic	Pleochroic	Pleochroic	Pleochroic
	Colour	Gray with brownish tint	Gray	Dark gray to lighter gray	Yellow-white	White to dark gray
	Reflectivity	Low	Low to moderate	Low	High	High
	Cleavage	Imperfect	Absent	Distinct	High	Perfect
	Polishing hardness	Very high	Low	Moderate	Very high	Very low

(Contd...)

	Property	Orpiment	Perovskite	Psilomelane	Pyrargyrite	Pyrite
Crossed nicol	Optic character	Isotropic	Anisotropic	Anisotropic	Anisotropic	Anisotropic
	Interference colour	–	Green	Bluish gray, violet gray	Green to greenish blue	White with a pink shade
	Twinning	Absent	–	Absent	Lamellar and interpenetrating	Lamellar
	Zoning	Pyrometasomatic varieties are zoned	Present	Absent	Absent	Indistinct
	Internal reflection	Absent	Deep green	Blood-red	Not visible	Absent
	Etch test	HCl turns yellow	–	$SnCl_2$ darkens	–	Negative
Mineral		Orpiment	Perovskite	Psilomelane	Pyrargyrite	Pyrite
Plane polarized light	Nature of polish	Poor	Good	Good	Good	Poor
	Shape and habit	Radiating, acicular	Euhedral or subhedral	Cryptocrystalline, amorphous, colloform	Subhedral	Euhedral
	Pleochroism	Nonpleochroic	Nonpleochroic	Nonpleochroic	Pleochroic	Nonpleochroic
	Colour	Gray-white	Gray	Gray-white	Gray-white with a bluish tinge	Light yellow
	Reflectivity	Low	Low	Moderate	Low	High
	Cleavage	Absent	Indistinct	Indistinct	Distinct	Indistinct
	Polishing hardness	Very low	Very high	Very high	Moderate	Very high
Crossed nicol	Optic character	Imperfectly anisotropic	Isotropic to weakly anisotropic	Anisotropic	Anisotropic	Isotropic, occasionally anisotropic
	Interference colour	Masked by internal reflection	–	Grayish white, brownish	Grayish to brownish tints	Yellowish to greenish tints
	Twinning	Absent	Lamellar	Absent	Lamellar	Absent
	Zoning	Absent	Absent	Absent	Occasionally present	Distinct
	Internal reflection	Light yellow	Orange-brown to yellowish	Absent	Red	Absent
	Etch test	–	–	Etching by H_2O_2	–	Strongly etched by HNO_3

(Contd...)

Mineral		Pyrolusite	Pyrrhotite	Realgar	Romanechite
Plane polarized light	Nature of polish	Poor	Excellent	Poor	Good
	Shape and habit	Reniform, colloform, cryptocrystalline	Subhedral, prismatic	Anhedral	Fibrous, colloform, cryptocrystalline
	Pleochroism	Pleochroic	Pleochroic	Nonpleochroic	Pleochroic
	Colour	Light yellow to light yellowish gray	Brownish creamy to reddish brown	Gray-white	Gray-white
	Reflectivity	High	Moderately high	Low	Moderate
	Cleavage	Imperfect, cracks frequent	Absent, parting present	Absent	Absent
	Polishing hardness	Very high	Moderately high	Very low	Moderately high
Crossed nicol	Optic character	Anisotropic	Anisotropic	Anisotropic	Anisotropic
	Interference colour	Bright yellow to yellow-brown	Yellowish gray greenish gray, bluish gray	Masked by internal reflection	Gray-white, bluish and brownish shades
	Twinning	Lamellar	Deformation twins	Uncommon	Uncommon
	Zoning	Concentric with psilomelane	May be present	Absent	Absent
	Internal reflection	Yellowish brown, greenish blue	Yellowish gray to bluish gray	Orange-red to orange-yellow	Not well marked
	Etch test	HNO_3 stains black	HNO_3 stains brown	–	–

Mineral		Siderite	Sphalerite	Stannite	Stibnite
Plane polarized light	Nature of polish	Good	Good	Good	Good
	Shape and habit	Anhedral to subhedral	Euhedral, tetrahedral	Anhedral to subhedral, frequently twinned	Anhedral to subhedral
	Pleochroism	Pleochroic	Nonpleochroic		Pleochroic
	Colour	Gray	Pale gray with brownish tint	Light gray with olive-green tint	White to gray
	Reflectivity	Low	Low	Moderate	High
	Cleavage	Distinct	Perfect	Imperfect	Imperfect
	Polishing hardness	Moderate	Low	Moderate	Low

(Contd...)

	Tenorite	Tetrahedrite-tennantite	Thorianite	Tungstenite	Twinnite
Crossed nicol					
Optic character	Anisotropic	Isotropic	Isotropic	Anisotropic	Anisotropic
Interference colour	Yellow, brown	–	–	Violet, blue	Bluish gray-white
Twinning	Occasional	Lamellar exhibited by etching	Lamellar	Lamellar	Lamellar and wedge shaped
Zoning	Uncommon	Common	Uncommon	Uncommon	Absent
Internal reflection	Yellow, brown	Reddish brown, yellowish brown	Red-brown	Violet, blue	Not seen
Etch test	–	Zonal texture and twinning may be developed by HNO_3	–	–	Aqua regia stains black
Plane polarized light					
Nature of polish	Good	Good	Good	Poor	Good
Shape and habit	Anhedral, acicular	Subhedral	Euhedral	Lamellar	Subhedral
Pleochroism	Pleochroic	Nonpleochroic	Nonpleochroic	Pleochroic	Pleochroic
Colour	Gray-white with brown tint	Gray-white, with greenish shade	Gray	White to gray	White
Reflectivity	Moderate	Moderate	Low	High	Moderate
Cleavage	Imperfect	Imperfect	Perfect	Micaceous	Imperfect
Polishing hardness	Moderate	Moderate	High	Low	Moderate

	Uraninite	Wolframite	Wurtzite	Zincite
Crossed nicol				
Optic character	Isotropic	Anisotropic	Anisotropic	Anisotropic
Interference colour	–	Gray, bluish	Yellow-white	Greenish gray, bluish
Twinning	–	Lamellar	Uncommon	Lamellar
Zoning	Frequent	Uncommon	Uncommon	Uncommon
Internal reflection	Red-brown	Indistinct	Rare	Red
Etch test	–	–	–	–
Plane polarized light				
Nature of polish	Good	Good	Good	Good
Shape and habit	Euhedral	Euhedral, lamellar	Fibrous	Anhedral
Pleochroism	Nonpleochroic	Weakly pleochroic	Nonpleochroic	Nonpleochroic
Colour	Gray	Light gray	Gray	Gray
Reflectivity	Low	Low	Low	Low
Cleavage	Distinct	Imperfect	Imperfect	Distinct
Polishing hardness	High	High	Moderate	Moderate

(Contd...)

	Optic character	Isotropic	Anisotropic	Anisotropic	Not discernable
	Interference colour	–	Brown to yellow-brown, greenish	Red, brown, yellow	–
Crossed nicol	Twinning	Uncommon	Uncommon	Uncommon	Uncommon
	Zoning	Uncommon	Occasional	Uncommon	Uncommon
	Internal reflection	Dark brown	Red	Red-brown, yellow-brown or yellowish	Red, orange, yellow
	Etch test	–	–	–	–

7

Determination of Chemical Formula of Mineral

Commonly a mineral is defined as a substance having a definite chemical composition formed by the inorganic processes of the nature. It has been observed that some minerals like calcite and hematite are secreted or formed by organisms that rules out the sole inorganic origin of minerals to certain extent. The chemical composition of a particular mineral species also varies within certain limits, which may not be appreciably significant. For example, the quartz, which is colourless in the purest form, occurs in different colours due to the presence of minor amounts of different elements in its crystal lattice. Similarly, sphalerite (ZnS), which is colourless in purest form, looks brown due to the presence of about 3.6% iron and black when the iron content increases up to 15.44% in addition to 0.3% cadmium and 1.01% lead in its crystal structure. There are many mineral groups, which have different species of contrasting chemical composition. For instance, the general formula of pyroxene group may be expressed as XYZ_2O_6, where X = Ca, Na; Y = Mg, Fe^{2+}, Fe^{3+}, Mn, Li, Al, Cr, Ti, etc; Z = Si, Al. This group constitutes series like enstatite - ferrosilite $(Mg, Fe^{2+})_2Si_2O_6$, diopside—hedenbergite-johannsenite $[Ca(Mg, Fe^{2+}, Mn) Si_2O_6]$, aegirine-augite $(Na, Fe^{3+}, Si_2O_6) - [(Na, Ca) (Fe^{3+}, Fe^{2+}, Mg) Si_2O_6]$ including species like spondumen $(LiAlSi_2O_6)$, jadeite $(NaAlSi_2O_6)$, etc. Substitution of one element for another and entrapment of elements of smaller ionic radii within the lattices constituted of bigger sized ions are common in mineral kingdom. Thus, the composition of a mineral, which is otherwise considered fixed, varies to certain extent as revealed by chemical analyses. The chemical composition of a mineral in terms of oxides and elements can be estimated by chemical analyses from which the formula of the mineral can be determined by a set of calculations. The approximate molecular weights of some common oxides and elements employed in computation are given in Table 7.1.

Before going for computation of mineral formula, let us analyse the case of common quartz having chemical formula SiO_2 and molecular weight 60. It contains one atom of silicon and two atoms of oxygen. The atomic weights of silicon and oxygen are 28 and 16 respectively. Out of 6 gram of SiO_2, silicon accounts for 28 gram and oxygen constitutes 32 gram. So the weight percentages of silicon and oxygen are 46.67 $\left(\dfrac{28}{60} \times 100\right)$ and 53.33 $\left(\dfrac{32}{60} \times 100\right)$ respectively.

The atomic proportions of silicon and oxygen are 46.67 ÷ 28 = 1.67 and 53.33 ÷ 16 = 3.33 respectively. If 1.67 is taken as one unit, the number of Si atom is 1.67 ÷ 1.67 = 1 and number of

Table 7.1: Atomic weight of common elements and molecular weights of common oxides used in calculation of mineral formulae

Oxide	Mol. wt.	Oxide	Mol. wt.	Oxide	Mol. wt.	Oxide	Mol. wt.
Al_2O_3	102	Cr_2O_3	152	MnO	71	SiO_2	60
B	11	CuO	79.6	MnO_2	87	SnO	134.7
B_2O_3	69.6	F	19	Na_2O	62	SrO	103.6
BaO	153.4	FeO	72	NiO	74.7	ThO_2	264
BeO	25	Fe_2O_3	159.7	P_2O_5	142	TiO_2	80
CO_2	44	H_2O	18	PbO	223.2	UO_2	270
CaO	56	K_2O	94.2	Rb_2O	187	V_2O_5	182
Cr_2O_3	152	Li_2O	30	S	32	ZnO	81.4
Cl	35.5	MgO	40.3	SO_3	80	ZrO_2	123.2

O atom is $3.33 \div 1.67 \approx 2$. The formula constitutes one Si atom and two O atoms making the mineral formula SiO_2.

The method of determination of chemical formula involves a set of calculations depending on the type of analysis or available data.

7.1 ELEMENTAL COMPOSITION OF THE MINERAL IS KNOWN

Step 1: Determine the atomic proportion (AP) of elements by dividing the weight percentages of different atoms by respective atomic weights.

Step 2: Take the average of nearly equal values of APs as unit (1). Determine the numbers of different ions constituting the mineral by dividing the APs of respective ions by the unit value and rounding the numbers to nearest integers. There may be some ions whose atomic proportions are much less compared to the unit AP. These are substituting ions, which occur in the crystal lattice substituting other ions or are primary ions of other mineral(s) present in minor amount.

Step 3: The formula of the mineral is determined from the number of primary constituent ions.

Step 4: Electrical neutrality of the mineral molecule is tested by algebraic sum of the valencies of constituent ions.

Example 1: The chemical analysis result of a sulphide mineral is given in Table 7.2. The formula is to be determined from which its name is to be ascertained.

Step 1: The percentages of Cu, Pb, Sb and S are 13.05, 42.87, 24.33 and 19.75. The corresponding atomic weights are 63.6, 207.2, 121.8 and 32.0 respectively. Determine the atomic proportions (APs) of the elements by dividing the weight percentages of different atoms by respective atomic weights. Atomic proportions of Cu, Pb, Sb and S are 0.205, 0.207, 0.2 and 0.615 respectively (Table 7.2).

Step 2: The nearly equal values are 0.205, 0.207 and 0.2. The average of these three numbers is 0.204. Determine the number of Cu, Pb, Sb and S ions by dividing respective atomic proportions by 0.204. The numbers of Cu, Pb, Sb and S ions rounded to nearest integers are 1, 1, 1 and 3 respectively.

Table 7.2: Calculation of number of different ions from the weight percent data of Example 1

Element/ion	Weight percent	Atomic weight	Atomic proportion (AP)	Number of ion(s)
Cu	13.05	63.6	$13.05 \div 63.6 = 0.205$	$0.205 \div 0.204 \approx 1.0$
Pb	42.87	207.2	$42.87 \div 207.2 = 0.207$	$0.207 \div 0.204 \approx 1.0$
Sb	24.33	121.8	$24.33 \div 121.8 = 0.2$	$0.200 \div 0.204 \approx 1.0$
S	19.75	32.0	$19.75 \div 32.0 = 0.615$	$0.615 \div 0.204 \approx 3.0$

Step 3: The formula of the mineral is $CuPbSbS_3$.

Step 4: Total electrical charge $= (+1) + (+2) + (+3) + (-2 \times 3) = 6 - 6 = 0$

The mineral molecule is electrically neutral. The formula suggests the mineral to be bournonite.

7.2 OXIDE COMPOSITION OF THE MINERAL IS KNOWN

Step 1: Determine the molecular proportion (MP) of oxides by dividing the weight percentages of different oxides by respective molecular weights.

Step 2: Take the average of nearly equal values of MPs as unit (1). Determine the numbers of different oxides constituting the mineral by dividing the MPs by the unit value and rounding the numbers to nearest integers. When the MPs show much variation, the value of intermediate MP is to be taken as unit (1) and numbers of different oxides are obtained by dividing the respective MPs by this unit value. There may be some oxides with much less values. These correspond to substituting ions, which occur in the crystal lattice substituting other ions of similar valency or size.

Step 3: The formula of the mineral is determined from the number of constituent oxides.

Step 4: Electrical neutrality of the mineral molecule (mineral formula) is tested by algebraic sum of the valencies of constituent ions.

Step 5: Necessary adjustments in the molecule (mineral formula) are made to account for substituting cations.

Example 2: The chemical analysis result of a mineral is given in Table 7.3. The formula is to be determined from which its name is to be ascertained.

Step 1: Determine the molecular proportions (MPs) of CaO, SO_3 and H_2O by dividing the weight percentages of oxides by respective molecular weights. These are 0.581, 0.581 and 1.163 respectively.

Step 2: The average of nearly equal values of MPs is 0.581, which is to be taken as 1 unit. Determine the numbers of all oxides by dividing the respective MPs by 0.581. Number of CaO, SO_3 and H_2O are 1, 1 and 2 respectively.

Table 7.3: Calculation of number of different oxides from the weight percent data of Example 2

Oxide	Weight percent	Molecular weight	Molecular proportion (MP)	Number of oxides
CaO	32.55	56	$32.55 \div 56 = 0.581$	$0.581 \div 0.581 = 1$
SO_3	46.51	80	$46.51 \div 80 = 0.581$	$0.581 \div 0.581 = 1$
H_2O	20.94	18	$20.94 \div 18 = 1.163$	$1.163 \div 0.581 = 2.001$

Step 3: The mineral formula is $CaO \cdot SO_3 \cdot 2H_2O$ or $CaSO_4 \cdot 2H_2O$.

Step 4: Total charge $= (+2) + (+6) + \{4 \times (-2)\} + [2 \times \{2 \times (+1) + (-2)\}] = 0$

The mineral molecule is electrically neutral. The formula suggests the mineral to be gypsum.

Example 3: The chemical analysis result of a mineral is given in Table 7.4. The formula is to be determined from which its name is to be ascertained.

Table 7.4: Calculation of number of different oxides from the wt. percent data of Example 3

Oxide	Weight percent	Molecular weight	Molecular proportion (MP)	Number of oxides
SiO_2	38.46	60	$38.46 \div 60 = 0.641$	$0.641 \div 0.213 = 3.01$
Al_2O_3	21.7	102	$21.7 \div 102 = 0.213$	$0.213 \div 0.213 = 1.00$
FeO	31.08	72	$31.08 \div 72 = 0.432$	$0.432 \div 0.213 = 2.03$
MnO	1.50	71	$1.50 \div 71 = 0.021$	$0.021 \div 0.213 = 0.10$
MgO	5.26	40.3	$5.26 \div 40.3 = 0.131$	$0.131 \div 0.213 = 0.62$
CaO	2.00	56	$2.00 \div 56 = 0.036$	$0.036 \div 0.213 = 0.17$

Step 1: Determine the molecular proportions (MPs) of SiO_2, Al_2O_3, FeO, MnO, MgO and CaO by dividing the weight percentages by respective molecular weights. These are 0.641, 0.213, 0.432, 0.021, 0.131 and 0.036 respectively.

Step 2: The intermediate MP, i.e. MP of Al_2O_3 is 0.213, which can be taken as equal to 1 unit. Determine the numbers of all oxides by dividing their MPs by 0.213. In this process the number of SiO_2 is determined to be $= 3.01 \approx 3$, $Al_2O_3 = 1$. FeO, MnO, MgO and CaO are all oxides of divalent cations. The sum of their numbers comes to be $2.03 + 0.1 + 0.62 + 0.17 = 2.92 \approx 3$.

Step 3: The mineral consists of 3 SiO_2, 1 Al_2O_3 and 3 (FeO, MnO, MgO and CaO). The formula is $3FeO.Al_2O_3.3SiO_2$, which is equivalent to $Fe_3Al_2(SiO_4)_3$. The formula suggests the mineral to be almandine garnet. Out of FeO, MnO, MgO and CaO, FeO has been taken in the mineral formula due to its predominance over others. Mn^{2+}, Mg^{2+} and Ca^{2+} are substituting ions.

Step 4: Total charge of the molecule is $(+2 \times 3) + (+3 \times 2) + 3 \times [+4 + (-2 \times 4)] = 12 - 12 = 0$

Step 5: MgO is the 4th dominant oxide (0.62). Replacement of Fe by Mg makes the formula $Mg_3Al_2(SiO_4)_3$ that corresponds to pyrope garnet. Similarly, MnO may be due to $Mn_3Al_2(SiO_4)_3$, i.e. spessartite garnet. CaO may be due to replacement of Fe, Mg and Mn by Ca in the crystal lattice. In conclusion, it can be said that the mineral whose analysis is given in Table 7.4 is dominantly almandine garnet with some amount of pyrope and spessartite.

7.3 OXIDE COMPOSITION OF THE MINERAL IN ADDITION TO H_2O, F, CL, ETC. ARE KNOWN

Step 1: Determine the molecular proportion (MP) of oxides by dividing the weight percentages by respective molecular weights.

Step 2: Determine the atomic proportion (AP) of oxygen from each oxide molecule by multiplying the molecular proportions (MP) by number of oxygen atoms present in the molecular formula. Sum up the atomic proportion of oxygen. Say it is equal to T.

Step 3: Determine the factor (F) by dividing the total number of oxygen (and equivalent radicals) present in mineral formula by T.

Step 4: Calculate the number of individual anions (N) present on the basis of total number of oxygen. These are obtained by multiplying respective atomic proportion of oxygen by F.

Step 5: Determine the number of individual ions present in the mineral formula by dividing N by number of oxygen per anion in respective oxide symbol.

Step 6: Inspect and add the number of ions to make the sums whole or nearly whole numbers.

Step 7: Determine the mineral formula by arranging anions to the left and cations to the right. The number of each anion will be the whole number found out in step-6. The number of oxygen and equivalent radicals will be the same as provided. These numbers should be shown by subscripts. Divide all the subscripts by their highest common factor, if necessary.

Step 8: Check the electrical neutrality by adding the positive and negative charges in mineral formula. The sum should be equal to zero.

Step 9: Keep the substituting anions and cations by the side of main anion and cation respectively separated by comma (,) within brackets.

Example 4: The chemical analysis result of a pyroxene is given in Table 7.5. The mineral formula is to be determined on the basis of 18 oxygen ions and the name of the pyroxene is to be ascertained.

Table 7.5: Calculation of number of different ions from the wt. percent data of Example 4

Oxide	SiO_2	FeO	MnO	CaO
Wt. percent	50.40	9.33	1.31	38.96
Mol. wt.	60.00	72.00	71.00	56.00
MP	0.84	0.130	0.018	0.696
No. of O	2	1	1	1
AP of O	1.680	0.130	0.018	0.696
F	7.132	7.132	7.132	7.132
N	11.982	0.927	0.128	4.964
Anion	Si	Fe	Mn	Ca
No. of O/ion	2	1	1	1
No. of ions	5.991	0.927	0.128	4.964

Step 1: Determine the molecular proportion (MP) by dividing the wt. percent by mol. wt.

Step 2: Determine the atomic proportion (AP) of oxygen by multiplying the molecular proportions (MP) by number of oxygen atoms present in the molecular formula. Sum up the atomic proportion of oxygen (T).

$$T = 1.68 + 0.13 + 0.018 + 0.696 = 2.524$$

Step 3: F = Total number of oxygen $\div$ 2.524 = 18 $\div$ 2.524 = 7.132

Step 4: N = AP of O $\times$ 7.132

Step 5: Determine the number of individual ions present in the mineral formula by dividing N by number of oxygen per anion in respective oxide symbol.

Step 6: Si = 11.982 $\div$ 2 = 5.991 $\approx$ 6;

Ca + Fe + Mn = 4.964 + 0.927 + 0.128 = 6.019 $\approx$ 6

Fe and Mn, whose sum is nearly equal to 1 has not been added with Ca because the general formula of pyroxene is XYZ_2O_6, in which number of X-ion = number of Y-ion. In case Ca and (Fe + Mn) are taken separately, number of X (Ca) will be 5 and number of Y (Fe + Mn) will be 1, which are not equal.

Step 7: The general formula of the mineral is $Ca_6Si_6O_{18}$. Highest common factor (HCF) of 6 and 18 is 6. So the mineral formula will be $CaSiO_3$.

Step 8: Charges of Ca = +2, Si = + 4 and O = –2.

Sum of charges = $(+2 \times 1) + (+ 4 \times 1) + (–2 \times 3) = +2 + 4 – 6 = 0$.

Step 9: In the calculation, Ca, Fe and Mn have been added up. So the final formula of the mineral is $(Ca, Fe, Mn)_2SiO_3$. The formula suggests the mineral to be wollastonite in which Fe and Mn have substituted Ca to certain extent.

Example 5: The chemical analysis result of an amphibole is given in Table 7.6. The mineral formula is to be determined on the basis of 24 oxygen ions and the name of the amphibole is to be ascertained.

Table 7.6: Calculation of number of different ions from the wt. percent data of Example 5

Oxide	SiO_2	Al_2O_3	Fe_2O_3	FeO	MgO	CaO	Na_2O	H_2O
Wt. percent	50.6	7.4	2.5	5.3	18	12.5	0.6	2.3
Mol. wt.	60	102	159.7	72	40.3	56	62	18
MP	0.843	0.073	0.016	0.074	0.447	0.223	0.010	0.128
No. of O	2	3	3	1	1	1	1	1
A.P. of O	1.686	0.219	0.048	0.074	0.447	0.223	0.01	0.128
F	8.466	8.466	8.466	8.466	8.466	8.466	8.466	8.466
N	14.274	1.854	0.406	0.626	3.784	1.888	0.085	1.084
Anion	Si	Al	Fe^{3+}	Fe^{2+}	Mg	Ca	Na	OH
No. of O/ion	2	1.5	1.5	1	1	1	0.5	0.5
No. of ions	7.14	1.24	0.27	0.63	3.78	1.89	0.17	2.17

Step 1: Molecular proportion (MP) is found out by dividing the weight percent by mol. wt.

Step 2: Atomic proportion (AP) of oxygen is determined by multiplying the molecular proportions (MP) by number of oxygen atoms present in the molecular formula. Sum up the atomic proportion of oxygen (T).

$$T = 1.686 + 0.219 + 0.048 + 0.074 + 0.447 + 0.223 + 0.01 + 0.128 = 2.835$$

Step 3: F = Total number of oxygen ÷ 2.835 = 24 ÷ 2.835 = 8.466

Step 4: N = AP of O × 8.466

Step 5: Determine the number of individual ions present in the mineral formula by dividing N by number of oxygen per anion in respective oxide symbol.

Step 6: The general formula of amphibole is $X_2Y_5Z_8O_{22}(OH)_2$

where, X = Ca, Na, Mg, Fe^{+2}, etc.; Y = Fe^{2+}, Fe^{3+}, Al, Mg, Mn, etc.; Z = Si, Al.

H_2O in chemical analysis is the water of crystallization that occurs as (OH) ion in mineral formula and takes the place of 2 oxygen ions. The number of OH = $2.17 \approx 2$

Ca + Na = 1.89 + 0.17 = 2.06 ≈ 2. Mg and Fe^{2+} should not be grouped with these

Si = 7.14, which can be made equal to 8 by allocating (8–7.14) = 0.86 of Al to it. This is justified because Al replaces Si in the crystal lattice.

The balance Al (0.38), Fe^{3+} (0.27), Fe^{2+} (0.63) and Mg (3.78) sum to 5.06 ≈ 5

Step 7: The general formula of the mineral is $Ca_2Mg_5Si_8O_{22}(OH)_2$.

Since the HCF of 2, 5, 8 and 22 is 1, the formula needs no change

Step 8: Charges of Ca = +2, Mg = +2, Si = + 4, O = –2 and OH = –1

Sum of charges = (+2 × 2) + (+2 × 5) + (+4 × 8) + (–2 × 22) + (–1 × 2)

$$= + 4 + 10 + 32 - 44 - 2 = 0.$$

Step 9: In the calculation Ca and Na; Mg, Fe^{3+}, Fe^{2+} and Al as well as Si and Al have been grouped. So the final formula of the mineral is:

$$(Ca, Na)_2 (Mg, Fe^{2+}, Al, Fe^{3+})_5 (Si,Al)_8 O_{22} (OH)_2$$

The formula suggests the mineral to be hornblende.

Note: OH is considered equivalent to O. In the chemical formula 2 OH and 22 O account for 24 oxygen ions given in the question.

Example 6: The chemical analysis result of a humite is given in Table 7.7. The mineral formula is to be determined on the basis of 18 oxygen ions and the name of the humite is to be ascertained.

Step 1: Determine the molecular proportion (MP) by dividing the wt. percent by mol. wt. H_2O^- is not taken into consideration because it does not occur in the crystal lattice.

Table 7.7: Calculation of number of different ions from the wt. percent data of Example 6

Oxide	SiO_2	Al_2O_3	Fe_2O_3	FeO	MgO	MnO	H_2O^+	H_2O^-	F
Wt. percent	36.8	0.2	0.5	4.2	54	0.3	1.2	0.1	2.7
Mol. wt.	60	102	159.7	72	40.3	71	18	–	19
MP	0.613	0.002	0.003	0.058	1.340	0.004	0.067	–	0.142
No. of O	2	3	3	1	1	1	1	–	–
AP of O	1.226	0.006	0.009	0.058	1.34	0.004	0.067	–	0.142
F	6.642	6.642	6.642	6.642	6.642	6.642	6.642	–	6.642
N	8.143	0.040	0.060	0.385	8.900	0.027	0.445	–	0.943
No. of O/ion	2	1.5	1.5	1	1	1	0.5	–	–
No. of ions	4.07	0.03	0.04	0.39	8.90	0.03	0.89	–	0.94

Step 2: Determine the atomic proportion (AP) of oxygen by multiplying the molecular proportions (MP) by number of oxygen atoms present in the molecular formula. Sum up the atomic proportion of oxygen (T).

$$T = 1.226 + 0.006 + 0.009 + 0.058 + 1.34 + 0.004 + 0.067 = 2.71$$

Note: F (0.142) has not been taken into consideration because O is not associated with it.

Step 3: F = 18 ÷ 2.71 = 6.642

Step 4: N = AP of O × 6.642

Step 5: Determine the number of individual ions present in the mineral formula by dividing N by number of oxygen per anion in respective oxide symbol.

Step 6: The general formula of humite is $Mg(OH, F)_2 \cdot (1-4) Mg_2SiO_4$

$$Mg + Fe^{3+} + Fe^{2+} + Mn = 8.9 + 0.04 + 0.39 + 0.03 = 9.36 \approx 9 = 1 + 8$$

$$OH + F = 0.89 + 0.94 = 1.83 \approx 2 \qquad Si + Al = 4.07 + 0.03 = 4.1 \approx 4$$

Step 7: The formula of the mineral is $Mg(OH, F)_2 \cdot 4(Mg_2SiO_4)$

Since the HCF of 2 and 1 is 1, the formula needs no change

Step 8: Charges of $Mg = +2, Si = +4, O = -2, OH = -1$ and $F = -1$

Sum of charges $= (+2) + (-1 \times 2) + 4[(+2 \times 2) + (+4) + (-2 \times 4)] = +2 - 2 + 4[+4 + 4 - 8] = 0$.

Step 9: In the calculation Mg, Fe^{+3}, Fe^{+2} and Mn; OH and F as well as Si and Al have been grouped. So the final formula of the mineral is

$$(Mg, Fe^{+3}, Fe^{+2}, Mn)(OH, F)_2 \cdot 4[(Mg, Fe^{+3}, Fe^{+2}, Mn)_2(Si, Al)O_4]$$

The formula suggests the mineral to be clinohumite.

Note: OH and F are considered equivalent to O. In the chemical formula 2(OH, F) and 16 O account for 18 oxygen ions given in the question.

7.4 PROBLEMS FOR SOLUTION

The chemical analyses (in weight percent) of 8 minerals are given in Table 7.8, 11 minerals are given in Table 7.9 and 11 minerals are given in Table 7.10. Determine the formula and suggest the name of the mineral in each case.

Table 7.8: Elemental composition data of minerals

Element	M-1	M-2	M-3	M-4	M-5	M-6	M-7	M-8
Fe	7.99	30.43	0.00	0.00	0.00	45.54	0.00	11.13
Cu	0.00	34.62	79.86	0.00	0.00	0.00	64.44	63.31
Pb	1.64	0.00	0.00	85.90	5.83	0.00	0.00	0.00
Zn	57.38	0.00	0.00	0.70	62.58	1.10	2.71	0.00
S	32.99	34.94	20.14	13.40	31.59	53.36	32.85	25.56

Table 7.9: Oxide composition data of minerals

Oxide/ element	M-9	M-10	M-11	M-12	M-13	M-14	M-15	M-16	M-17	M-18	M-19
SiO_2	34.4	38.6	36.6	36.3	48.0	48.4	63.5	64.0	40.2	59.4	48.20
TiO_2	0.4	0.5	0.0	0.0	0.1	0.8	0.0	0.0	0.1	0.0	0.0
B_2O_3	0.0	0.0	0.0	10.2	0.0	0.0	0.0	0.0	0.0	0.0	0.0
Al_2O_3	1.1	18.2	62.6	40.3	11.0	27.0	19.5	22.3	32.5	25.8	0.40
Fe_2O_3	1.5	5.6	0.4	0.00	0.6	6.6	0.1	0.1	2.0	0.6	2.04
FeO	40.5	3.8	0.1	3.3	1.6	0.8	0.0	0.2	0.0	0.0	23.50
MnO	0.8	0.7	0.0	1.0	0.0	0.0	0.0	0.0	0.0	0.0	3.50
MgO	20.5	0.8	0.1	0.1	20.0	0.0	0.0	0.3	0.1	0.0	1.00
CaO	0.8	31.7	0.2	0.7	12.4	0.0	0.5	3.2	1.7	0.1	21.36
Na_2O	0.0	0.0	0.0	2.1	2.4	0.4	0.8	9.8	11.0	13.6	0.0
K_2O	0.0	0.0	0.0	0.5	1.2	11.2	15.6	0.1	12.4	0.1	0.0
Li_2O	0.0	0.0	0.0	1.2	0.0	0.0	0.0	0.0	0.0	0.0	0.0
H_2O^+	0.0	0.1	0.0	4.0	0.8	4.3	0.0	0.0	0.0	0.4	0.0
F	0.0	0.00	0.0	0.3	1.9	0.5	0.0	0.0	0.0	0.0	0.0
No. of O	4	24	20	31	24	24	32	32	32	6	6

Table 7.10: Oxide composition data of minerals

Oxide/ element	M-20	M-21	M-22	M-23	M-24	M-25	M-26	M-27	M-28	M-29	M-30
TiO_2	0.0	19.5	0.0	0.0	0.0	0.0	0.0	0.0	0.0	0.0	0.0
Al_2O_3	0.2	1.6	10.3	0.0	0.0	0.0	0.0	0.0	0.0	0.0	0.0
Cr_2O_3	0.0	0.1	59.5	0.0	0.0	0.0	0.0	0.0	0.0	0.0	0.0
Fe_2O_3	68.9	28.8	3.3	0.0	0.0	0.0	0.0	0.0	0.0	0.0	0.0
FeO	30.8	47.2	14.1	0.0	0.6	0.3	58.8	0.40	0.0	0.0	0.0
MnO	0.1	0.4	0.2	0.0	0.2	60.9	2.9	0.0	0.0	0.0	0.0
MgO	0.0	2.3	12.6	0.2	46.9	0.0	0.1	21.50	0.9	0.0	0.0
CaO	0.0	0.1	0.0	55.9	0.4	0.5	0.1	30.40	52.2	55.8	54.5
CO_2	0.0	0.0	0.0	43.9	51.9	38.3	38.1	47.70	0.0	0.0	0.0
P_2O_5	0.0	0.0	0.0	0.0	0.0	0.0	0.0	0.0	40.2	42.4	41.8
H_2O^+	0.0	0.0	0.0	0.0	0.0	0.0	0.0	0.0	0.0	1.8	0.0
F	0.0	0.0	0.0	0.0	0.0	0.0	0.0	0.0	0.00	0.0	3.7
Cl	0.0	0.0	0.0	0.0	0.0	0.0	0.0	0.0	6.7	0.0	0.0
No. of O	32	32	32	6	6	6	6	6	26	26	26

7.5 ANSWERS OF PROBLEMS GIVEN IN SECTION 7.4

M-1: $(Zn, Fe)S$—Sphalerite

M-2: $CuFeS_2$—Chalcopyrite

M-3: Cu_2S— Chalcocite

M- 4: $(Pb, Zn)S$—Galena

M-5: $(Zn, Pb)S$—Sphalerite

M-6: $(Fe, Zn)S_2$—Pyrite

M-7: $(Cu, Zn)S$—Covellite

M-8: Cu_5FeS_4—Bornite

M-9: $(Mg, Fe^{+2})_2SiO_4$—Hortonolite (Olivine)

M-10: $(Ca, Fe^{+2})_3(Al, Fe^{+3})_2Si_3O_{12}$—Grossular + Andradite (Garnet)

M-11: Al_2SiO_5—Andalusite

M-12: $(Na, Fe^{+2}, Mn, Ca) (Li, Al)_3 B_3Si_6O_{27} (OH, F)_4$—Elbaite (Tourmaline)

M-13: $(Na, K) Ca_2Mg_5Si_7AlO_{22} (OH, F)_2$—Edenite + Pargasite (Hornblende)

M-14: $(K, Na)_2 (Al, Fe^{+3})_4 (Si, Al)_8 O_{20} (OH, F)_4$—Muscovite

M-15: $(K, Na) (Al, Si) Si_3O_8$—Orthoclase (Alkali Feldspar)

M-16: $(Na, Ca) Al (Si, Al)_3O_8$—Oligoclase (Plagioclase Feldspar)

M-17: $Na_3 (Na, K, Ca) (Al, Fe^{+3})_4Si_4O_{16}$—Nepheline (Feldspathoid)

M-18: $NaAlSi_2O_6$—Jadeite (Pyroxene)

M-19: $CaFeSi_2O_6$—Hedenbergite (Pyroxene)

M-20: $Fe^{+2}Fe_2^{+3}O_4$—Magnetite

M-21: $(Fe^{+2}, Fe_2^{+3}, Ti, Mg)O_4$—Titaniferous magnetite

M-22: $(Mg, Fe^{+2})(Cr, Al, Fe^{+3})_2O_4$—Magnesian chromite

M-23: $CaCO_3$—Calcite

M-24: $MgCO_3$—Magnesite

M-25: $MnCO_3$—Rhodochrosite

M-26: $FeCO_3$—Siderite

M-27: $CaMg(CO_3)_2$—Dolomite

M-28: $Ca_5(PO_4)_3Cl$—Chlor-apatite

M-29: $Ca_5(PO_4)_3(OH)$—Hydroxy-apatite

M-30: $Ca_5(PO_4)_3F$—Fluor-apatite

Bibliography

1. Dana, ES and Ford, WE (1959). A textbook of mineralogy. Asia Publishing House, New Delhi.
2. Deer, WA, Howie, RA and Zusman, J (1978). An introduction to the rock forming minerals. ELBS and Longman, London.
3. Flint, E (1971). Essentials of crystallography. Mir Publishers, Moscow.
4. Hurlbut, CS (Jr.) and Klein, C (1977). Manual of mineralogy. John Wiley and Sons, New York.
5. Mitra, S (1989). Fundamentals of optical, spectroscopic and X-ray mineralogy. Wiley Eastern Ltd., New Delhi.
6. Perkins, D (2002). Mineralogy. Prentice-Hall of India Pvt. Ltd. New Delhi.
7. Read, HH (1984). Rutley's Elements of Mineralogy (26th Ed.). CBS Publishers and Distributors, New Delhi.
8. Sen, AK (1995). Laboratory manual of Geology. Modern Book Agency Pvt., Ltd, Kolkata.
9. Whitten, DG A and Brooks, JRV (1972). The Penguin dictionary of Geology. Penguin Books Ltd., Harmondsworth.

Index